11/12/93

The Care and Conservation of Geological Material: Minerals, Rocks, Meteorites and Lunar Finds

Butterworth-Heinemann Series in Conservation and Museology

The Care and Conservation of Geological Material:
Minerals, Rocks, Meteorites and Lunar Finds

Edited by
Frank M. Howie

Butterworth-Heinemann Ltd
Linacre House, Jordan Hill, Oxford OX2 8DP

 PART OF REED INTERNATIONAL BOOKS

OXFORD   LONDON   BOSTON
MUNICH   NEW DELHI   SINGAPORE   SYDNEY
TOKYO   TORONTO   WELLINGTON

First published 1992

© Butterworth-Heinemann Ltd 1992

British Library Cataloguing in Publication Data
Howie, F.M.
    Care and Conservation of Geological
    Material: Minerals, Rocks, Meteorites and
    Lunar Finds. – (Conservation and Museology Series)
    I. Title   II. Series
    551.0288

ISBN 0 7506 0371 2

Library of Congress Cataloguing in Publication Data
The Care and Conservation of Geological Material: Minerals, Rocks,
    Meteorites, and Lunar Finds/edited by Frank M. Howie.
    p.   cm. – (Butterworth–Heinemann series in conservation and museology)
    Includes bibliographical references and index.
    ISBN 0 7506 0371 2
    1. Minerals – Collection and preservation. 2. Rocks – Collection
    and preservation. 3. Meteorites – Collection and preservation.
    4. Lunar petrology. I. Howie, F. (Frank) II. Series.
    QE366.2.C37   1992   91–45839
    552'.075–dc20   CIP

Composition by Scribe Design, Gillingham, Kent
Printed and bound in Great Britain

Series editors' preface

The conservation of artefacts and buildings has a long history, but the positive emergence of conservation as a profession can be said to date from the foundation of the International Institute for the Conservation of Museum Objects (IIC) in 1950 (the last two words of the title being later changed to Historic and Artistic Works) and the appearance soon after in 1952 of its journal *Studies in Conservation*. The role of the conservator as distinct from those of the restorer and the scientist had been emerging during the 1930s with a focal point in the Fogg Art Museum, Harvard University, which published the precursor to *Studies in Conservation, Technical Studies in the Field of the Fine Arts* (1932–42).

UNESCO, through its Cultural Heritage Division and its publications, had always taken a positive role in conservation and the foundation, under its auspices, of the International Centre for the Study of the Preservation and the Restoration of Cultural Property (ICCROM), in Rome, was a further advance. The Centre was established in 1959 with the aims of advising internationally on conservation problems, co-ordinating conservation activators and establishing standards of training courses.

A significant confirmation of professional progress was the transformation at New York in 1966 of the two committees of the International Council of Museums (ICOM), one curatorial on the Care of Paintings (founded in 1949) and the other mainly scientific (founded in the mid-1950s) into the ICOM Committee for Conservation.

Following the Second International Congress of Architects in Venice in 1964 when the Venice Charter was promulgated, the International Council of Monuments and Sites (ICOMOS) was set up in 1965 to deal with archaeological, architectural and town planning questions, to schedule monuments and sites and to monitor relevant legislation.

From the early 1960s onwards, international congresses (and the literature emerging from them) held by IIC, ICOM, ICOMOS and ICCROM not only advanced the subject in its various technical specializations but also emphasized the cohesion of conservators and their subject as an interdisciplinary profession.

The use of the term *Conservation* in the title of this series refers to the whole subject of the care and treatment of valuable artefacts both movable and immovable, but within the discipline conservation has a meaning which is distinct from that of restoration. *Conservation* used in this specialized sense has two aspects: first, the control of the environment to minimize the decay of artefacts and materials; and, second, their treatment to arrest decay and to stabilize them where possible against further deterioration. Restoration is the continuation of the latter process, when conservation treatment is thought to be insufficient, to the extent of reinstating an object, without falsification, to a condition in which it can be exhibited.

In the field of conservation conflicts of values on aesthetic, historical, or technical grounds are often inevitable. Rival attitudes and methods inevitably arise in a subject which is still developing and at the core of these differences there is often a deficiency of technical knowledge. That is one of the principle *raisons d'être* of this series. In most of these matters ethical principles are the subject of much discussion, and generalizations cannot easily cover (say) buildings, furniture, easel paintings and waterlogged wooden objects.

A rigid, universally agreed principle is that all treatment should be adequately documented. There is also general agreement that structural and decorative falsification should be avoided. In addition there are three other principles which, unless there are overriding objections, it is generally agreed should be followed.

The first is the principle of the reversibility of processes, which states that a treatment should normally be such that the artefact can, if desired, be returned to its pre-treatment condition even after a long lapse of time. This principle is impossible to apply in some cases, for example where the survival of an artefact may depend upon an irreversible process. The second, intrinsic to the whole subject, is that as far as possible decayed parts of an artefact should be conserved and not replaced. The third is that the consequences of the ageing of the original materials (for example 'patina') should not normally be disguised or removed. This includes a secondary proviso that later accretions should not be retained under the false guide of natural patina.

The authors of the volumes in this series give their views on these matters, where relevant, with reference to the types of material within their scope. They take into account the differences in approach to artefacts of essentially artistic significance and to those in which the interest is primarily historical, archaeological or scientific.

The volumes are unified by a systematic and balanced presentation of theoretical and practical material with, where necessary, an objective comparison of different methods and approaches. A balance has also been maintained between the fine (and decorative) arts, archaeology and architecture in those cases where the respective branches of the subject have common ground, for example in the treatment of stone and glass and in the control of the museum environment. Since the publication of the first volume it has been decided to include within the series related monographs and technical studies. To reflect this enlargement of its scope the series has been renamed the Butterworth-Heinemann Series in Conservation and Museology.

Though necessarily different in details of organization and treatment (to fit the particular requirements of the subject) each volume has the same general standard which is that of such training courses as those of the University of London Institute of Archaeology, The Victoria and Albert Museum, the Conservation Center, New York University, the Institute of Advanced Architectural Studies, York, and ICCROM.

The authors have been chosen from among the acknowledged experts in each field, but as a result of the wide areas of knowledge and technique covered even by the specialized volumes in this series, in many instances multi-authorship has been necessary.

With the existence of IIC, ICOM, ICOMOS and ICCROM, the principles and practice of conservation have become as internationalized as the problems. The collaboration of Consultant Editors will help to ensure that the practices discussed in this series will be applicable throughout the world.

Contents

Foreword

I recall the time, some 15 years ago, when the realization came to me that a substantial part of the collections under my care were not terribly useful for the kinds of research that they were meant to serve. They had altered considerably during the years they had rested on the shelves of the museum and, in some instances, the specimens were nothing more than a record of a particular species at a point in time and space. In a few cases the labels had become far more important than the specimens they documented. My fellow curators were aware of the problem but, like me, they did not know how to deal with it. I began to look beyond the jars of specimens in alcohol and discovered that mineralogists had even greater problems – some of their specimens were no longer what they were when they were collected.

Most people (and many biologists) assumed that a mineral was a mineral and that was the end of that. I was soon disabused of that simplistic point of view, and I was shown some of the subtle changes, and occasionally some that were not so subtle, going on in the collections that were 'solid as a rock'. I was even more surprised to learn that these changes were being studied by a few dedicated scientists in a few institutions and that they were beginning to understand the effects that the museum environment had on their specimens. It comes as no surprise to me to see their names included as authors in this text. Some of these people had even gone to the 'extreme' of altering the environment around their specimens to minimize the changes. For me, these were revolutionary approaches and I followed them with great interest and began to practice some of the principles that seemed applicable to my specimens.

The conservation of natural history collections is really no different to the conservation of paintings or ethnographic collections. It is a matter of understanding the materials you are trying to conserve, how they are affected by the environment they are in and attempting to provide an environment that will have a minimal effect on them. This book goes far toward the goal of understanding the materials that are found in mineral collections and characterizing the problems that arise. It provides a critical examination of the effects of temperature, light and humidity on minerals; these after all are the same environmental problems that plague every curator in every collection of every type all over the world. The chemical changes that occur in mineral collections are examined in detail and methods for their prevention or, at least, reducing the rate of change are discussed.

This text is not a compendium of recipes to cure the ills of every mineral collection in the museums of the world. Rather, it is an explanation of the problems encountered by those who have to deal with the long-term preservation of some of the most exquisite creations to be found on this planet and their relatives that have made their way here with, and without, the help of technology developed by humankind. It seems to me terribly fitting that we are beginning to express our concerns for their preservation and that the authors of this text have seen the need to tell the rest of us how we can begin to accomplish this goal. I have no doubt that some of the ideas expressed in this work will change over time; in fact I am sure that the authors, and others, have already begun the research that will change our present views during the next few decades. That is the norm for the advancement of knowledge. For the present, however, we can rest assured that we have a sound foundation on which to base decisions for the conservation of our mineral collections and we owe a debt of gratitude to the authors of this text for have provided us with the facts on which to base those decisions.

C. G. Gruchy
Director, Canadian Conservation Institute
Ottawa

ix

Preface

This volume has been compiled to help fill one of the many gaps in the literature on natural history collection conservation. Although it is primarily intended for those with responsibilities for the care of institutional geological collections, it is hoped that it will be of assistance to both private collectors and students in museology.

The book has developed from a sketchy idea of the editor, thanks to the knowledge and patience of the contributors, into an attempt at a 'state of the art' text. It is hoped that it will encourage others to tackle many areas where problems continue to cause loss and decay of the diminishing resource of natural materials.

Since materials, processes and techniques used in the conservation of specimens and artifacts alter and evolve on a continuous basis, it was decided from the outset to emphasize more the principles of care through the identification and explanation of basic mineral instabilities rather than describe specialized treatment methods for preservation. Armed with the information contained in this book, not only should the conservation practitioner be able to devise treatments specific to the problems in hand, but the curator, collection manager or collector should be able to improve standards of care.

The data given on mineral stability is as accurate as possible, given the present state of knowledge on the subject. At the time of publication there is considerable renewed interest in many aspects of mineral stability, sparked off in large measure by the concern for natural resource conservation and environmental issues.

Sincere thanks are offered to the contributors for their time, expertise and forbearance in the preparation of their chapters. To Monica Price (Stability of minerals); to Kurt Nassau (Conserving light-sensitive minerals and gems); to Rob Waller (Temperature and humidity sensitive mineralogical and petrological specimens); to Alex Bevan (Meteorites); to Charles Meyer (The Lunar sample collection) and to Bob King for Appendices II, III and IV. The editor is also much indebted to the many colleagues who have reviewed the manuscript and offered many helpful suggestions. The responsibility for any errors in the text is of course mine alone.

FRANK HOWIE

Notes on Contributors

Alex Bevan is Curator of Minerals in the Western Australian Museum, Perth, Australia. He was previously a scientist in the Department of Mineralogy of the Natural History Museum in London. His interests lie in the development of curatorial methods for mineral and petrological collections and research into meteorites and other extraterrestrial material.

Frank M. Howie is Safety and Conservation Advisor at the Natural History Museum in London. He is also a consultant on natural history conservation and museum safety. His interests lie in the areas of surface properties of sulphides, development of contaminant control systems and raising the profile of conservation research in the natural history field.

Bob King is Curator of the John Moore Museum in Tewkesbury. He was previously Curator of Rocks and Minerals at the National Museum of Wales, Cardiff. He is editor of the *Journal of The Russell Society*. His interests are in mineral collecting, processing and curation techniques.

Charles Meyer is Associate Lunar Curator for NASA at the Lyndon B Johnson Space Center in Houston, Texas. His chief interests are lunar sample contamination control, dating zircons by ion microprobe mass analysis and planning lunar sample studies.

Kurt Nassau recently retired as Distinguished Research Scientist from AT&T Bell Laboratories in Murray Hill, New Jersey. He has also taught at Princeton University. With many years of experience in academic and industrial research he has published extensively on the physics and chemistry of solid state materials, including crystal chemistry, crystal growth, colour, minerals, and gems. He is now a consultant and author.

Monica Price is Assistant Curator of the Mineral Collections at the University Museum in Oxford. She has been a member of the Geological Curators' Group Committee and has published mineralogical contributions for recent collection care guidelines.

Robert Waller is Head, Conservation Section at the Museum of Nature in Ottawa where he was previously a conservation scientist in the Mineral Sciences Division. He has also worked in the Canadian Conservation Institute on mineral stability projects. His interests are in the conservation of minerals and of fluid-preserved biological collections. He has been involved as a consultant in many North American natural history museums since 1986 and with the planning and the implementation of collections care pilot training programme.

Introduction

Of the 3,500 or so known mineral species, perhaps only 350 are inherently unstable. All specimens or samples of minerals and rocks, however, whether stable or not, deserve the same high degree of care, as each will add to the sum of scientific knowledge.

The conservation of minerals must of necessity start in the field. Many of the unstable species require immediate protection from vibration and water vapour. Most importantly, adequate location data should accompany the mineral or rock sample from the moment it is collected. This data may be as important as the specimen itself, and without it, the specimen's scientific value is considerably debased. Adequate care should be taken while processing material for study or exhibition.

There is extensive literature concerned with processing mineral specimens, however not all of it has the best interest of the specimen in mind. Any treatment of mineral specimens, however minor, has the potential for changing physical or chemical properties, perhaps permanently. Some methods have been developed for just such a reason, for example the heat treatment and chemical dyeing of gem minerals and the modification or removal of accessory minerals to enhance the specimens' appearance or value.

The curation of mineral and rock specimens presents special problems, and there is the need to develop a specific methodology. Mineral specimens often have extremely high financial value, and the security of collections has often to be considered a high priority. Handling specimens with hazardous properties can give rise to problems, and radioactive species require special storage and handling techniques.

The obligations of a curator of rare, consumable but irreplaceable materials such as meteorites have been admirably reviewed by Max Hey in his article entitled 'The Curator's Dilemma', published in 1969. The basic tenets of this contribution are worth restating, and of course are applicable to all collected specimens:

> On the one hand, he (the curator) must make material available for research and study: on the other, he must preserve material in his care for enjoyment and study by future generations. The two requirements are diametrically opposed, and it is therefore his duty to work out a reasoned compromise.

Particular problems have long existed with the storage and exhibition of environmentally susceptible minerals. Many of these problems have not been treated as problems requiring scientific examination. Light-sensitive minerals, efflorescent and deliquescent species and reactive sulphides have specific environmental requirements which are only now being quantified.

It is hoped that the approach to mineral conservation embodied in this volume will assist and encourage all those interested and involved in the preservation of the geological heritage, whether as students, professionals working in a major institution or purely as mineral collectors.

1

The stability of minerals

Monica Price

It has long been recognized that all kinds of objects and artifacts are liable to deteriorate if stored in less than ideal conditions. Paper, ivory and wood will shrink and crack if allowed to dry out too much, or warp and develop mould growths in humid conditions. Many metals will tarnish or corrode; pigments on cloth, paper, paint and animal skins will all too readily fade on continued exposure to light; and all manner of plant and animal material will succumb to the ravages of pest insects.

At first sight, geological specimens – rocks, minerals and fossils – might seem to be remarkably durable. But closer examination reveals that they are by no means so resilient. In the earth's crust minerals form in equilibrium with their environment. But the earth is a mobile, constantly changing environment for the minerals and rocks enclosed within it. Rocks are crushed and melted by major tectonic forces, or dissolved, oxidized and eroded by water and air. Old minerals are destroyed, new ones form. In spite of this, the majority of rocks and minerals are surprisingly durable, but a significant minority, including some common species, will alter or perish in our normal living environment.

1.1 The development of mineralogy

The roots of modern mineralogy lie in the studies of metallurgy and medicine, and many of the first collections were established to illustrate courses in these subjects in the European mining academies and older universities. As a pure science, mineralogy developed most rapidly in the nineteenth century, hand in hand with contemporary developments in chemical theory, and launched by René Just Haüy's classic research into crystal symmetry as a basis for mineral classification.[1] Interest in the science was matched by a rise in the popularity of mineral collecting as a pastime. Now, some instabilities and their remedies were readily discernible, e.g. realgar, proustite and pyrargyrite would only retain their pristine colours if they were kept in the dark. Other metastabilities were noted but remained unexplained.[2]

It was not until the first half of the twentieth century that the deteriorating conditions of geological specimens in museum collections was to arouse widespread professional concern. In 1922 A.L. Parsons of the University of Toronto presented a paper to the Mineralogical Society of America, describing how a number of minerals required specific environmental conditions - darkness, low or high humidity, steady or low temperature, and avoidance of exposure to radioactivity - if they are to be preserved.[3] This aroused a lively response, and four years later Parsons listed a further selection of vulnerable minerals.[4] In the 1930s F.A. Bannister described the solutions applied by the British Museum (Natural History) to the stability problems of marcasite and pyrite,[5] and to a range of other minerals and meteorites.[6]

Decomposing specimens were, undoubtedly, inconvenient for the researcher, but perhaps only then was the concept of specimens as historic records of their localities acquiring widespread recognition. Each specimen provides a unique material representation of the geology of a particular geographic location at a specific point in time. In Victorian times collectors and curators were rarely tolerant excess duplication of specimens, and all too many collections were further depleted through total neglect during the early years of the twentieth century. Storage conditions; pollution from open fires, industry and smogs, and large seasonal fluctuations in temperature and humidity were equally destructive. Much of the surviving geological collections in museums was rapidly becoming irreplaceable.

The loss of mineral localities continues. Uneconomic mines and quarries are often flooded or infilled, and political changes have barred access to many localities; for example, Soviet and Eastern European localities were, in practice, inaccessible to western collectors for several decades prior to the 1990s. Opportunities for collecting are being further restricted where minerals are designated 'cultural property' and subjected to export controls. Increasingly, new mining methods which crush ore underground permit fewer undamaged mineral specimens to reach the surface. Furthermore, specimen acquisition is an expensive business. Field collecting costs both time and money, and fine specimens, once freely given to museums, are acquiring the market and value normally attributed to works of art.

1.2 The function of mineral collections

Mineral and rock collections provide a valuable research tool and reference function essential to the needs of educational institutions and industry. Minerals are naturally formed chemical compounds, each 'species' being defined by a chemical composition which varies only between narrow limits, and an orderly arrangement of atoms which may be manifested by flat crystal faces. Rocks may rarely be monomineralic; most are composed of a mixture of minerals.

Systematic mineralogy is fundamental to mineralogical research. The designation of 'new' minerals is a process strictly governed by the International Mineralogical Association's Commission on New Minerals and Mineral Names. Around 3,500 species are recognized, approximately 55 to 80 more are validated each year.[7] The specimen or specimens upon which the description of a new species is based are designated 'type' specimens. As in the life sciences, mineralogical 'types' are the cornerstones of the science. Their preservation in 'non-private, institutional, professionally-curated, research-orientated museums'[8] is essential. All specimens upon which published scientific research has been based also require preservation in perpetuity in suitable institutions.[9,10]

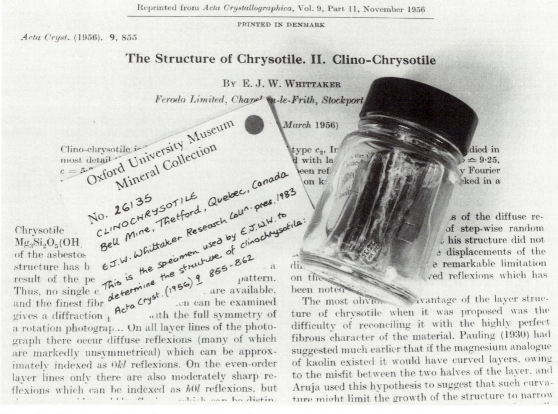

Figure 1.1 Clinochrysotile. Samples upon which published mineralogical research is based must be carefully documented and preserved in suitable scientifically-orientated institutions if they are to be accessible to future research workers.

Our understanding of earth history and processes is derived as much from detailed analyses of rock and mineral specimens as from the study of geological formations and structures. Observation of polymorphism, order–disorder transformations and exsolution in minerals can give a detailed thermal history of the host rock, while absolute dating can be achieved by measurement of radioactive isotopes present in certain minerals. The study of mineral paragenesis and topographic distribution is an essential process in the search for new ore deposits, and can yield new mineral species. Rocks, cores, polished ore samples and thin sections provide detailed records of geological formations both for reference and research, and for the training of future professional geologists.

Mineral collections have important uses outside the field of geology. They are essential for comparisons between synthetic compounds and their natural analogues, for theoretical research and a wide range of industrial applications, not least environmental research.

1.3 Mineral diversity, formation and stability

There are about 80 chemical elements (excluding short-lived products of radioactive decay) naturally occurring in the crust of the earth. Of these, eight – oxygen, silicon, aluminium, iron, calcium, sodium, potassium and magnesium – constitute almost 99% of the crust, by far the most common being oxygen, with 47% by weight and over 90% by volume.[11] This abundance of oxygen leads to a wide diversity of mineral species on earth, compared with, say, those found on the fresh, unweathered lunar surface. Besides forming simple oxides, oxygen forms strong bonds with carbon, phosphorus, silicon, etc., making the oxyanions which characterize the carbonate, phosphate and silicate groups of minerals. Furthermore, many minerals vary from one another simply in degree of hydration.

Water is important, not only as a constituent of minerals but also as a powerful medium for their creation and alteration. At moderate temperatures and atmospheric pressure minerals may be dissolved, or crystallize out, through the action of meteoric groundwater. At high temperatures and pressures saline aqueous solutions become powerful solvents, capable of dissolving even the least soluble silicate minerals.

It is the chemical elements present, and the bonds that hold them together, which dictate the chemical and physical properties of each mineral. Most elements have distinct preferences as to the elements or radicals to which they will bond, governed by their electron configuration. The 'siderophile' elements will preferentially remain in an elemental state or form natural alloys. They include gold and the platinum group elements, and are characterized by a remarkable chemical stability. The 'chalcophiles' prefer to form covalent bonds with sulphur, selenium or tellurium. They include cobalt, nickel, zinc, silver, lead and mercury. The 'lithophiles' are the largest group, including the alkalis, alkali earths, rare earth elements, halogens, carbon, phosphorus, silicon, manganese and iron. They most readily bond with oxygen to form oxides or oxyanions.

Such preferences will be modified by the absence of the preferred geochemical environment. Thus iron, a lithophile element, will readily bond with sulphur in a reducing environment. The resultant iron sulphide, pyrite, marcasite or pyrrhotite will convert to the sulphate in damp oxidizing conditions, releasing sulphuric acid which will, in turn, react with many associated species, converting them to sulphates. In the absence of both oxygen and sulphur, iron will form elemental native iron. Compared with such siderophiles as gold and platinum, native iron is both extremely rare and chemically unstable, and will readily oxidize at its contact with the atmosphere.

Some elements are reasonably abundant but do not form minerals. They are thinly dispersed as trace elements, substituting for ions of similar dimension and charge. Gallium, for example, is only found substituting for aluminium in aluminosilicates, and rubidium replaces potassium in potassium-bearing minerals.

Stable phases of most minerals form only under specific conditions of temperature, pressure and chemical concentration. If conditions deviate from the 'stability field' of one species, another species which is stable in the new environment will preferentially crystallize out.

Once formed, a species remains stable so long as the environment remains within its stability field. Outside that field the species becomes metastable. If sufficient energy is applied to the system, it will change to a more stable product; with no energy input, it will continue to exist in an apparently stable manner. In practice, the transition to a more stable polymorph, or the unmixing of more complex minerals into distinct and more stable constituent species, does take place, but so slowly that it only becomes apparent with the increasing maturity of a rock on a geological timescale.

The rarest minerals are characterized by the most restricted stability fields; the more common minerals tend to form in a broad range of geological environments encompassing a wide range of temperature, pressure and chemical parameters. Clearly the environment of formation of a mineral gives some guidance as to its potential for instability.

1.3.1 The stability of primary minerals

Igneous minerals, crystallized out from magmatic melts, are dominated by silica and certain groups of silicate minerals – the feldspars, feldspathoids, pyroxenes, amphiboles, micas and members of the olivine group. Accessory minerals may include titanite, zircon, and the rare platinum group metals, as well as certain oxides – chromite, magnetite and ilmenite. The high density of some accessories may lead to gravity settling and concentration into valuable economic deposits. If fractional crystallization of the magma takes place, late-stage concentration of rare elements and a supply of fluorine- and boron-bearing volatiles permit the formation of an additional assemblage of minerals typifying the pegmatitic environment – tourmaline, apatite, fluorite, beryl, topaz, wolframite, a range of metallic sulphides and oxides, including cassiterite and those of the rare earth elements. The minerals formed depend on the nature of the magma, which may be enriched by leaching of the country rock. Many are also formed under pneumatolytic and hydrothermal conditions: crystallization from hot gases and waters enriched with dissolved elements. Primary hydrothermal minerals also include the zeolites, most metallic elements and sulphides, many oxides, carbonates, fluorite, and the least soluble of the sulphates – barite and celestine.

Fine crystals are commonly formed in these environments, which are among the most important for the collector. Most primary igneous minerals are relatively hard and durable, and chemically resilient, as is demonstrated by their presence in many clastic sediments. Subtle changes continue to take place with increasing maturity of the rock, e.g. the transition from a high to low temperature phase in the quartz/tridymite/cristobalite system, or chemical unmixing of feldspars to form perthites. However, the chemical breakdown of the less stable igneous minerals, such as feldspars to clays, will only take place through exposure to the more powerful geological agents or over a prolonged timescale.

By comparison, the primary pegmatitic, pneumatolytic and hydrothermal minerals are more variable in both their physical and chemical stabilities. Silicates such as beryl and topaz, and many oxides, are durable; apatite, fluorite, the sulphates and carbonates are more vulnerable to abrasion and cleavage, and also to chemical attack. The sulphides and related minerals vary widely in physical durability between, for example, hard pyrite and soft molybdenite, and many are particularly susceptible to chemical oxidation. The zeolites are an exceptional group of silicate minerals in terms of their facilities for reversible dehydration and the base exchange of a wide range of cations. While fibrous zeolites are notable for their extreme fragility, all are vulnerable to chemical alteration.

The principal constituents of metamorphic rocks include many of the more stable groups of minerals formed in igneous environments: silica, the feldspars, pyroxenes, amphiboles, olivine group and micas, as well as the softer sheet silicates; the serpentines, chlorites and clays, and various carbonates. A wide range of oxides and sulphides also occur as accessory minerals. Certain other minerals are characteristic of metamorphic environments, and, in very general terms, the highest temperature/pressure conditions, with associated metamorphic dehydration, generate the more robust and chemically resilient species, such as corundum, sapphirine and spinel. The ortho- and di-silicates – kyanite, andalusite, staurolite, chloritoid, garnet, zoisite, epidote and cordierite – are generally robust, while brucite, talc and graphite are markedly soft.

Low temperatures and pressures characterize sedimentary environments. Most medium- and coarse-grained clastic rocks by definition comprise the more resilient components of their parent rocks, but fine-grained clastic rocks are generally composed of silica, the clays and/or carbonates. Where oxygen is depleted, organic material is readily preserved, and sulphides such as pyrite may be formed by the action of anaerobic bacteria. This increases the potential for destabilization under oxidizing conditions. Bioclastic rocks are composed predominantly of the carbonate minerals calcite, aragonite and dolomite, all moderately soft and vulnerable to a greater or lesser degree to attack by acid solutions. Primary phosphate deposits, typically bone beds, are dominated by members of the apatite group, but guano deposits, especially in caves, are characterized by less stable hydrous phosphates, readily soluble even in dilute acids.

The most unstable groups of minerals – the sulphates, halides, nitrates, iodates, and certain borates and carbonates – are most commonly formed by evaporation of saline waters, or by fumarolic activity (deposition from hot springs). Chemical variation is caused by the ratio of dissolved elements, seasonal temperature fluctuations, and such additional factors as, for example, local igneous activity introducing boron to the environment. Soft, and easily abraded, evaporite minerals are, by definition, water-soluble to a greater or lesser degree. They may be vulnerable to hydration, dehydration, efflorescence (the spontaneous loss of water) or deliquescence (the spontaneous absorption of water). Even opaline silica formed under fumarolic conditions, though comparatively hard, is prone to dehydration and cracking (see Chapter 4).

1.3.2 Secondary minerals

The majority of mineral species are of secondary origin. Beneath the water-table anaerobic conditions exist, and groundwater becomes an agent for the formation of secondary sulphides. Above the water-table percolating meteoric groundwater, often acidic, leaches and oxidizes native elements, sulphides, arsenides, etc., to their secondary salts – the sulphates, carbonates, phosphates, arsenates, and vanadates. A variety of secondary silicates may also be formed. Secondary alteration is common in mines and mine dumps, and different assemblages in which halides are prevalent are formed by the action of seawater on primary and secondary mineral deposits. Secondary salts exhibit wide variations in degrees of hydration and this, along with the multiple valency of some of the most common elements, e.g. iron and manganese, results in a remarkable diversity of species. Well-formed crystals are common, but they are rarely robust, and show restricted stability, being altered or destroyed by prolonged exposure to chemical solutions.

1.3.3 Extraterrestrial minerals

The most exotic mineral environment is that of outer space. Here it is the lack of oxygen, and hence water, that results in a strictly limited range of mineral species, an absence of secondary alteration; but, more specifically, the formation of a number of species which are not known in terrestrial environments. Extraterrestrial minerals reach the earth either through man's sampling of the lunar environment or through the natural collision of meteoritic material with the earth. Certain meteoritic minerals are notoriously unstable. Lawrencite (iron chloride) readily oxidizes in air to form iron hydroxide and hydrochloric acid, the latter attacking associated species, particularly meteoric iron. Oldhamite (calcium sulphide) will react with water to form gypsum. The preservation of extraterrestrial material in a pristine state often requires isolation in an anoxic environment (see Chapters 6 and 7).

1.4 Mineral stability, processing and storage

Minerals vary immensely in their stability and response to environmental changes, even in the earth's crust. As soon as a mineral specimen is recovered from its terrestrial setting and transferred to a storeroom, it is brought into a new environment, which serves to protect it from certain natural agencies; extremes of temperature and pressure, solution, biological action and prolonged radiation. In turn it may, however, be exposed to new agencies, such as fluctuating temperatures and humidities, light, mechanical abrasion due to vibration and handling, atmospheric dust and pollution, and the introduction of chemicals in labelling and storage materials, conservation and development treatments. Some reactions may be rapid, e.g. the change of colour of light-sensitive species in bright sunlight; others are slow and only perceptible over a period of years.

Research into the metastabilities of minerals has moved hand in hand with the development of chemical theories, and recent research, notably by Howie,[12] Waller[13,14] and Nassau,[15] has shown that understanding of processes enables both prediction and prevention of damage.

1.5 Mineral conservation: basic principles

In terms of mineral stability, conservation can be of two kinds: preventive conservation, where alteration and damage is ideally forestalled (but more often reduced to the slowest possible rate); and remedial conservation, which is the reversal or repair of damage already done. Clearly for mineral specimens as natural chemical compounds there is very limited scope for remedial conservation. Foreign matter, dust and dirt may be cleaned off, but it is usually only possible to arrest the process of chemical decay, and then at the risk of introducing new, artificial chemicals which may alter the natural physical features of a species – its colour, lustre or surface texture – or totally destroy associated species. Only in exceptional circumstances, in some light-activated colour changes or rehydration of certain dehydrated minerals, is alteration truly reversible. Preventive conservation is therefore especially important.

But the concept of conservation must be taken beyond that of merely providing suitable environmental conditions for the storage of specimens. Each sample bears associated documentation – where it was collected, by whom, and when – data which is primary and potentially irretrievable if lost. Conservation must apply not only to the whole specimen but to its associated data. Fundamentally it encompasses all the processes which provide both for the long-term preservation of geological specimens, and for their subsequent use in display, education and research.

1.5.1 Mineral conservation and the collector

Conservation begins in the hands of the collector, even before a specimen is removed from the ground. Conservation-conscious collectors seek to understand the locality they collect from, to recognize and appreciate the value of the material they collect, and to ensure that by respecting the legal rights of landownders, site managers and mineral-rights

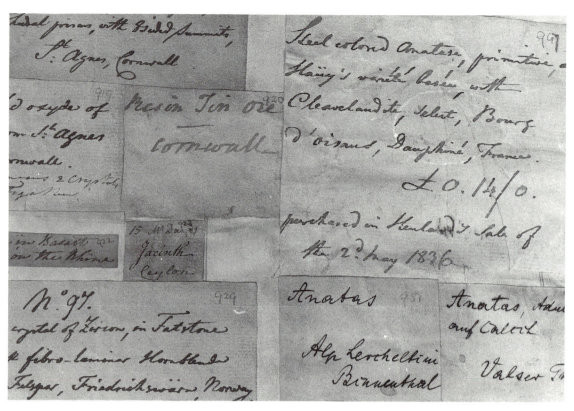

Figure 1.2 Original labels. Without its associated data, a specimen is virtually worthless. Here original labels are stored in albums, and form an important historical collection in their own right. They are protected from damage by unstable or soft, friable minerals, but are linked to them by registration number. Modern museum labels accompany the specimens in storage.

owners, and by their behaviour in the field, they do nothing that will discredit the mineral-collecting community or that might bar access to other collectors.

Most rock specimens are collected for research, reference or educational use, but minerals are acquired for a wider variety of reasons. The academic mineralogist, company geologist or museum curator will be primarily concerned to extract scientific information from the specimen, whether for educational or research purposes. This interest is shared by an increasing body of amateur collectors, traditionally the stalwarts of topographic mineralogy. Geological students of all ages seek reference material relevant to their studies, while for many other collectors the goal is the finest, most aesthetically pleasing, or rarest specimen. It may be acquired by exchange or purchase, but is especially prized when 'self-collected'. The dealer, earning a living by collecting, seeks a market across the whole spectrum of the mineral community, and must find

material appropriate to the needs of scientist and aesthete alike.

Although collecting by professional geologists is normally directed to satisfy specific research interests, for many hobbyists it starts out as a random affair, seeking all specimens and from any localities. The normal progression shows our collector becoming more discerning as to the aesthetic quality or rarity of his specimens, and encountering that age-old problem of restricted storage space. The former is a healthy progression. By familiarizing himself with the localities he is to visit and the species it may yield, he can be more selective about his specimens, and know their inherent instabilities. He will learn to study his specimens carefully for tiny or fragile crystals, and use tools and containers which will best protect samples from sudden environmental change or mechanical abrasion. This familiarization comes in part through books, but works most effectively through the study of specimens in private collections and public museums. Equally valuable is the guidance of other, more experienced collectors, and the coordinated activities of amateur clubs and societies.

The problem of lack of storage space has a number of solutions. Many collectors start to collect smaller and smaller samples – 'thumbnails' and 'micros'. Such is the nature of mineral specimens that it is easier to find small fine specimens than larger ones, and indeed many species only occur as microcrystals. Immense satisfaction is derived from the collecting of small specimens, their subsequent trimming and mounting, although doubts have been expressed by Dunn and Francis[16] about the increasing market in tiny samples of rare species. Each species occurs as part of an assemblage of minerals, and over-zealous trimming all too often results in the loss of one or more components of that assemblage, with a corresponding reduction in scientific value.[17]

The other, popular, solution is for collectors to specialize in a favourite group of minerals, or in specimens from a particular geographic locality – a country, district, mine or quarry. Undoubtedly such specialization, accompanied by high standards of documentation, results in the formation of important and valuable reference collections.

The importance of accurately recorded field data cannot be overemphasized; the precise locality, geological environment, date of collection and name of collector comprise vital records. Unless this information is promptly and honestly recorded, it is vulnerable to the failure of human memory. Dubious or incorrect data reduce the scientific value of a specimen.[18] The deliberate distortion of field data by certain collectors and dealers in the past has resulted in the squandering of valuable curatorial time, and the confusion or discrediting of research results.

1.5.2 Mineral conservation and the curator

It would be impossible to consider the part the collector has to play in the conservation of geological specimens in isolation from that of the curator. Their roles are generally shared; most professional curators participate in primary collecting activities, while most collectors take on the role of curator through the care of their own personal collections. Clearly it is desirable that collectors and curators should share common high standards in the care of geological specimens.

One of the most contentious issues of common interest to collector and curator alike is the ethics of mechanical and chemical treatments which enhance the appearance of mineral specimens for aesthetic purposes. Dunn, Bentley and Wilson list a range of processes which range from genuine 'repairs' to outrageous 'fakes'.[19] They include simple repairs of crystals, restoration using plastics, the polishing or

etching of crystal faces to improve lustre, application of mineral oils to improve clarity, colour-enhancement processes and many other procedures which alter the natural appearance of the specimen. The dividing line between repair and fake is a narrow one. Even more contentious is the subject of mineral 'cleaning' or 'development'. Clearly the removal of foreign matter – clay, soil and dust – from a specimen after collection, or as remedial treatment following poor storage, is a necessary curatorial procedure. Tackled intelligently, with due regard to the fragility and solubility of all species present, it should leave the entire mineral assemblage undamaged.

However, 'cleaning' has also been interpreted to mean the removal of one or more less attractive mineral species – chlorite, iron or manganese oxides and hydroxides, and other secondary species. King[20] lists the main arguments against such procedures. He warns of the destruction of one or more components of a natural association of mineral species, with a consequential loss of scientific interest in that association, the creation of an etched or unnatural anaemic appearance, resulting in the loss of scientific and often aesthetic value, and the fact that no two mineral associations respond to a chemical reaction in the same way. He also states that the use of chemicals may produce a chemical imbalance which may be difficult or even impossible to neutralize, and that the removal of chemicals used in a development technique, both chemically and physically, can produce additional long-term risk to mineralogical material, labels and the immediate storage area. To these might be added a sixth argument, that many of the chemicals advocated for such procedures are extremely hazardous to the user. Few collectors have access to suitable facilities, or the expertise to use the strong acids, alkalis and other powerful reagents recommended in various 'recipes'.

Obviously there are circumstances where development of mineral specimens, analogous to that of palaeontological material, is a valid or essential procedure. Here, only by removing one species can another be revealed for study. Sympathetic development requires the use of the least damaging technique available, permits partial retention of the obscuring mineral to retain the complete association, and, most importantly, requires that full and careful documentation of the procedure be retained with the specimen. It is regrettable that there is no tradition or convention among dealers or collectors for indicating on labels the cleaning and development procedures which have been used.

The specimen market is lively. Samples that have passed through the hands of many owners are snapped up with as much alacrity as those fresh from the mine. Whatever the initial motivation for collecting a specimen may have been, it is likely to have encountered a number of careful, or less than careful, owners. Only the experience of handling a wide range of mineral specimens can teach the curator to recognize the frauds, fakes and enhanced specimens of the mineral world; and, even then, too many treatments which destroy scientific value leave ambiguous or barely detectable traces.

The initial care of newly accessed material is critical. Documentation procedures - the numbering, labelling and registration of specimens - must be both thorough and safe. All the paper, glue, card and other materials used for documentation and storage should be those least likely to endanger the chemical stability of minerals. The storage environment should be clean, dust-proof, and environmentally controlled to as high a standard as the resources of the individual or institution permit. Radioactive and toxic minerals should be stored in a manner most conducive to the conservation of other specimens, the curator, associated workers and visitors.[21,22]

But even in store, specimens require regular attention, and may deteriorate if neglected. The decomposition of the least stable species may initiate a cascade of destructive chemical reactions destroying both specimens and accompanying documentation. Routine inspection will reveal the onset of metastable reactions and allow remedial action to be taken before associated species and labels are damaged.

It is fortunate that most rocks and the majority of mineral species will remain acceptably stable under certain minimum environmental standards: a dry, preferably dark store, dust-proof containers, and minimum fluctuations in temperature and humidity from a 50% rh/15-20°C ideal. The creation of carefully monitored microclimates, regulated by the use of an agent such as conditioned silica gel, will extend the range of species which can be stored. But the more exacting requirements of the most metastable species, especially when they are scientifically important specimens or reference suites, can only be provided by institutions with the political will and financial resources to provide a good geological conservation service. It is no credit to individuals or institutions that they should retain material that they do not have the expertise or resources to preserve.

Good curatorial standards are detailed by Brunton, Besterman and Cooper,[23] and Knell and Taylor have written a valuable guide[24] for non-geologists in charge of geological collections. King has written a series of papers explaining basic collecting and curatorial techniques, aimed at the amateur curator,[25,26,27] who may also encounter sporadic papers in *The Mineralogical Record* and *The Journal of the Russell Society* describing new or improved techniques for the conservation and care of collections.

A wise collector, who has taken the trouble to look after his collection well, will plan well in advance for its long-term safekeeping. What will happen if his interest wavers, illness or death strikes? Good collections command high financial returns. There may be some rare and exceptional specimens, but all too often the true value of a collection lies in its whole, rather than the individual parts. It is a real loss to science when a collection representative of a particular area, collected over many years, is split up and sold off in lots.

If a collector wishes to ensure a permanent home for his collection in a museum, he should ensure first that the collection is appropriate to the needs and scope of the museum he has chosen, and that it has both staff and resources to safeguard the collection on a *permanent* basis. Judicious disposal of mineral specimens by collectors is matched by an equal need for judicious acquisition by curators. The curator

Figure 1.3 Rock store. Overcrowded and unsorted specimens, crates and boxes in a dark basement lacking environmental monitoring or climate control - the typical scenario for damage and decay in rock and mineral specimens. Decades of neglect have left many museum stores in conditions far worse than this example.

Figure 1.4 Student 'erosion'. Examples of typical student 'erosion': scratches from hardness tests, abrasion from careless handling, and misappropriation of specimens. The use of fine or rare specimens in teaching requires strict supervision.

must decide whether he[28] has the facilities to accommodate specimens offered to him, whether they are adequately documented, and whether they are a genuinely useful addition to his institution. Often he must weigh historical importance against shortfalls in documentation or quality.

Can he guarantee the long-term safekeeping of the material; the time, materials and manpower to catalogue and clean the collection; a suitable environment in which to store it; good security against theft, fire or natural disasters; wise restrictions on its use and disposal? There is a growing movement among public museums to publish acquisition policies which specify exactly what material a museum is committed to acquiring and preserving. Any curator drafting such a policy would be wise to bear in mind the conservation requirements of material he may be offered, as well as the more traditional criteria – type of object and geographic provenance.

1.6 Conclusions

It would be pointless to collect rocks and minerals without the intention of using them. Usage often takes place at the expense of conservation. The display of specimens may expose them to light levels and fluctuating temperatures/humidities which may alter or destroy some species. Damage to a very minor component of a specimen may jeopardize the stability of the whole. The vulnerability of a specimen to damage in a display environment and its importance in an undamaged state must be weighed against its value in a display.

Specimens used in academic teaching are vulnerable to attrition ('student erosion'), consequent upon less than expert handling and injudicious use of hardness and acid testing. If scientifically important specimens are to be used then strict supervision is essential. Some minerals should never be used for teaching purposes.

Most scientifically important material finds its greatest use in academic research. The organization of the collection must permit easy access to individual specimens; and for a collection to be used to its full potential, it is necessary to publish details of holdings. Much geological research requires the use of a tiny fragment of the parent specimen, whether for destructive or non-destructive techniques. It often falls to the curator to assess the relative value of the research or the appropriateness of the technique, and hence the expendability of the specimen. The curator must be a wise judge in the disposal of the material in his care.

Many of the best curated collections are owned by amateur collectors who have a thorough knowledge of the material in their care, and value its importance to our mineralogical heritage as well as its remarkable aesthetic appeal. Equally there are too many obsessive collectors whose sole satisfaction is the thrill of the hunt; they may destroy localities, and yet cannot make time to sort through, study, document and conserve their findings. Problems are by no means confined to collectors. Too many geological collections in public institutions do not have the care of a professional geological curator,[29] and too many geological curators lack the broad experience or facilities to conserve the very wide range of materials encountered in their collection. Professional geological conservators are still a rare breed. However, public museums and their staff can set both examples and standards in conservation. By working together, collectors and curators can predict and prevent damage to an important and irreplaceable part of our geological heritage.

References

1　HAÜY, R.J., '*Traité de mineralogie*' Paris, (1801)
2　DANA, E.S., '*The system of mineralogy of James Dwight Dana*', 6th ed. John Wiley & Sons, New York, pp. 34, 96 (1892)
3　PARSONS, A.L., 'The preservation of mineral specimens', *Amer. Min.*, 7, pp. 59–63 (1922)
4　PARSONS, A.L., 'Additional data concerning the preservation of mineral specimens', *Amer. Min.*, 11, pp. 79–83 (1926)
5　BANNISTER, F.A., 'The preservation of pyrites and marcasite', *Museums Journal*, 33, pp. 72–4 (1933)
6　BANNISTER, F.A., 'The preservation of minerals and meteorites', *Museums Journal*, 36, pp. 465–76 (1937)
7　DUNN, P.J. and MANDARINO, J.A., 'The Commission on New Minerals and Mineral Names of the International Mineralogical Association; its history, purpose and general practice', *Min. Record*, 19 pp. 319–23 (1988)
8　DUNN, P.J. and MANDARINO, J.A., 'Formal definitions of type mineral specimens', *Min. Mag.* 52, pp. 129–30 (1988)
9　GEOLOGICAL CURATORS' GROUP, 'Recommendations for the preservation of research collections', *The Geological Curator*, 3, pp. 290–91 (1982)
10　DUNN, P.J., 'Protocols for scientists on the deposition of investigated mineral specimens', *Min. Record*, 19, p. 291 (1988)
11　BERRY, L.G. and MASON, B., '*Mineralogy: concepts, descriptions, determinations*', 2nd ed., revised by R.V. Dietrich W.H. Freeman & Co., San Francisco (1983)
12　HOWIE, F.M., 'Museum climatology and the conservation of palaeontological material', *Special Papers in Palaeontology*, 22, pp. 103–25 (1979)
13　WALLER, R., 'The preservation of mineral specimens', *Proceedings, Amer. Inst. of Conservation*, 8th Annual Meeting, pp. 116–28 (1980)
14　WALLER, R., 'The prevention of deliquescence, efflorescence and hydration in mineral specimens', *7th Triennial meeting of ICOM Committee for Conservation*, 84.13.8–84.13.10 (1984)
15　NASSAU, K., '*Gemstone enhancement*' Butterworths, London (1984)
16　DUNN, P.J. and FRANCIS, C.A., 'Dangers to science from species dealers, *Min. Record*, 17, p. 226 (1986)
17　WILSON, W.E., DUNN, P.J. and BENTLEY, R.E., 'Mineral specimen trimming', *Min. Record*, 17, pp. 163–5 (1986)
18　BENTLEY, R.E., WILSON, W.E. and DUNN, P.J., 'Mineral specimen mislabelling', *Min. Record*, 17, pp. 99–103 (1986)
19　DUNN, P.J., BENTLEY, R.E. and WILSON, W.E., 'Mineral Fakes', *Min. Record*, 12, pp. 197–219 (1981)
20　KING, R.J., 'The care of minerals, Section 2 - the development of minerals', *J. Russell Society*, 1(2), pp. 54–77 (1983)
21　HENDERSON, P., 'Hazards in the curation and display of mineral and rock specimens with especial emphasis on radioactivity', *The Geological Curator*, 3, pp. 292–6 (1982)
22　PUFFER, J.H., 'Toxic Minerals', *Min. Record*, 11, pp. 5–11 (1980)
23　BRUNTON, C.H.C., BESTERMAN, T.P. and COOPER, J.A., 'Guidelines for the curation of geological materials', *Geol. Soc. Misc. Papers*, 17 (1985)
24.　KNELL, S.J. and TAYLOR, M.A., 'Geology and the local museum' HMSO, London (1989)
25　KING, R.J., 'Section 1-the cleaning of minerals', *J. Russell Society*, 1(1), pp. 42–53 (1982)
26　KING, R.J., 'The care of minerals. Section 3-the curation of minerals', *J. Russell Society*, 1(3), pp. 94–112 (1985)
27　KING, R.J., 'The care of minerals. Section 3B-the curation of minerals', *J. Russell Society*, 1(4), pp. 129–48 (1986)
28　WHITE, J.S., 'Some aspects of modern mineral collection curation', *Min. Record*, 22, pp. 251–4 (1991)
29　DOUGHTY, P.S., 'The state and status of geology in U.K. Museums', Misc. Pap. Geol. Soc. London, 13, 118 pp. (1981)

2

Conserving light sensitive minerals and gems

Kurt Nassau

In a discussion on the preservation of art objects it was pointed out that:

> . . .the ideal environment. . .would consist of an unilluminated region with cool constant temperature and a low constant humidity. When objects are stored in such an environment, as has happened in Egyptian tombs with their treasures and in European caves with their wall paintings, thousands of years can pass without significant deterioration; unfortunately, such conditions do not permit us to view the objects![1]

To the curator of art and artifacts incorporating organic dyes or organic vehicles containing pigments, this is a dominant consideration in his activities, while the curator of minerals and gems has, on the face of it, much less serious problems. The surprisingly extended listings in the Tables 2.3, 2.4 and 2.5 and the equally astonishing lack of adequate reviews in the literature nevertheless indicate that such complacency cannot be justified.

For an extended discussion on this subject, refer to Pearl,[2,3] and King.[4] These authors rely heavily on older work by Parsons[5] and Bannister.[6] There is a partial listing of light-sensitive minerals by Sinkankas[7] and there are relevant sections in Nassau.[1,8] There are undoubtedly many pertinent observations in the voluminous descriptions of individual minerals from varied localities in the literature, yet there appears to be no direct way to locate such data within a reasonable timeframe. The writer will welcome comments on significant omissions and errors from readers.

2.1 The effects of light on minerals and gems

It is an unfortunate paradox that the colours we admire are the result of the absorption of light and that we obviously require light to observe these colours; but it is exactly this absorbed light that can produce fading and other photochemical degradations, so that, in time, the colour and even the specimen itself can be completely destroyed. In this connection one usually includes the action of the highly active ultraviolet radiation band immediately adjacent to the visible region, since some of this always accompanies visible light in the absence of careful filtering.

2.1.1 Light, colour and alterations

These can be broken down into three slightly overlapping groups. First, there are light-induced colour changes without any other physical or chemical changes; these may or may not be reversible. Second, there are those light-induced alterations producing significant bulk physical or chemical changes. Last, there are light-activated surface reactions with air, moisture, and/or pollutants. The last two effects are inevitably irreversible.

It cannot be overemphasized that just as certain minerals occur in a variety of colours based on the presence of different impurities, so the behaviour of a specific substance on light exposure can similarly vary widely. Thus some brown topaz specimens are stable to light, while others fade rapidly;[8] similarly, while the colour of most smoky quartz[9] and amethyst is perfectly stable, some rare specimens do fade on exposure to light. Here again, differences in impurities or other imperfections are the cause. Many of the colour-causing mechanisms involved in colour alterations are not well understood and others are under active investigation. Just as in the past, these interpretations can be expected to change with time; unfortunately, the phenomena themselves do not change!

2.1.2 Light and energy

Light is that part of the electromagnetic spectrum that can be detected by our eyes. The colours of the spectrum can be directly related to one of a number of direct or inverted energy scales, three of which are given in Figure 2.1. These energy quantities are those contained in a photon, which is the smallest unit of light that can exist by itself. If radiation such as ultraviolet or infrared is present, the more general term quantum is used.

There is an important advantage in using a direct energy scale in electron volts (eV) (on the left in Figure 2.1), because photon or quantum values can be added or subtracted. If, say, an orange light photon with energy 2 eV is absorbed by an atom, then the atom's energy will be raised by 2 eV; if two such photons are absorbed, then 4 eV is added, and so on.

The higher the energy of a photon or quantum, the more likely it is that an alteration can be produced. Thus blue light will usually be more photo-active or actinic than red light, and the ultra-violet radiation usually accompanying visible light is even more active. This is important in the museum environment, where direct daylight is particularly damaging and where incandescent lamps are superior to fluorescent lamps and even to heavily filtered daylight, as shown in Table 2.1.

Table 2.1 The amount of damaging ultraviolet radiation in equal quantities of light

Illumination	Relative damage factor
Vertical skylight, open	100
Vertical skylight, window glass	34
Vertical skylight, UV-absorbing plexiglass	9
Fluorescent lamp	9
Incandescent lamp	3

Infrared radiation is, however, not completely harmless, since it may cause localized heating, particularly in direct sunlight. This can produce total destruction by shattering of a very heat-sensitive material, such as crystals of sulphur. Essentially all chemical reactions are accelerated by heat, including reactions with moisture and oxygen. The effect of the heat is to excite bonds or atoms, thus providing additional energy which is then available to initiate or to accelerate the reaction. When photons are absorbed, they can similarly produce local excitations and initiate or accelerate specific reactions, in addition to the more general acceleration produced indirectly via a heating effect.

2.1.3 Light absorption, emission and causes of colour

Quantum theory explains the absorption and emission of light at the atomic level. As one example, the Cr^{3+} ion that gives the red colour to corundum Al_2O_3 in ruby can have only discrete energy levels, and energy is absorbed or emitted during transitions between these levels (the positions of the levels are dependent on the 'ligand field' and the transitions are controlled by 'selection rules'). Thus a ruby in the normal 'ground state' $^4A_2$ in the energy level diagram at the left in Figure 2.2 can absorb either violet or green light quanta from a beam of white light, with excitation of the chromium ions into the $^4T_1$ or $^4T_2$ energy levels respectively. The remaining non-absorbed light passing through the ruby is red with a little blue, as shown at the right in Figure 2.2. The $^4T_1$ and $^4T_2$ state Cr^{3+} ions now emit some of their excess energy as heat to the ruby crystal while falling down to the ^2E state, and then emit the rest of the excess energy while returning to the ground state with the emission of the red ruby fluorescence at 1.79 eV, as shown in Figure 2.2. In moving from Cr^{3+} - containing ruby, to Cr^{3+} - containing emerald, there are changes in the position of the $^4T_1$ and $^4T_2$ levels, resulting in a green colour; since the ^2E energy level does not change, the fluorescence remains red. For further details see Nassau.[1]

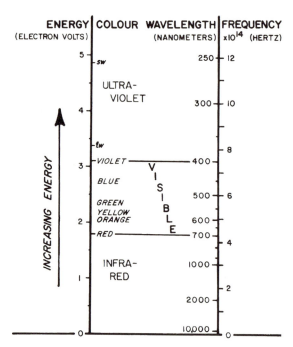

Figure 2.1 The spectrum, with three ways of numerically specifying the spectral colours.

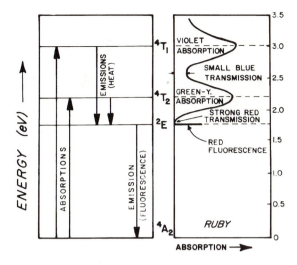

Figure 2.2 The energy levels and transitions in ruby (left), and the resulting absorption spectrum and fluorescence of ruby (right).

There are no less than 15 physical and chemical mechanisms which can produce colour, and these are summarized in Table 2.2. All but the first two of these mechanisms occur in minerals and gems. Colour changes produced by light are most common in the colour centres (cause 11 in Table 2); the processes involved are discussed in the next section. There are some additional colour alterations derived from transition metal impurity valence changes, either in ligand field (cause 5) or charge transfer (cause 7) environments, and some based on organic compounds (cause 6); all of these are discussed in Section 23, together with some general comments on photochemical changes. Detailed listings of specific photosensitive minerals are given in Tables 2.3, 2.4 and 2.5, and recommendations for protection are given in Section 2.5.

2.2 The nature of colour centres

When a smoky quartz crystal is heated to about 400°C for one hour or so, it will lose its colour. If it is next exposed to an intense source of energetic radiation, such as cobalt-60 gamma rays, then within a few minutes the smoky colour begins to reappear and the crystal will darken over some hours to a much deeper colour than the original. The colour of smoky quartz derives from a colour centre, as do the colours in amethyst, blue topaz, and irradiated diamonds. Some colour centres, such as all the ones mentioned so far, are usually perfectly stable, fading only when heated. Other colour centres, such as those in some brown topaz, in Maxixe beryl, and in

Table 2.2 Examples of the fifteen causes of colour

Vibrations and simple excitations
1 Incandescence: flames, lamps, carbon arc, limelight.
2 Gas excitations: vapour lamps, lightning, auroras, some lasers.
3 Vibrations and rotations: water, ice, iodine, blue gas flame.

Transitions involving ligand field effects
4 Transition metal compounds: turquoise, malachite, chrome green, some fluorescence, lasers, and phosphors.
5 Transition metal impurities: ruby, emerald, aquamarine, red iron ore, some fluorescence and lasers.

Transition between molecular orbitals
6 Organic compounds: most dyes, most biological colorations, some fluorescence and lasers.
7 Charge transfer: blue sapphire, magnetite, lapis lazuli, vivianite, ultramarine, Prussian blue.

Transitions involving energy bands
8 Metals: copper, silver, gold, iron, brass, 'ruby' glass.
9 Pure semiconductors: silicon, galena, cinnabar, sulphur, greenockite, realgar, diamond.
10 Doped semiconductors: blue and yellow diamond, light-emitting diodes, some lasers and phosphors.
11 Colour centres: amethyst, smoky quartz, desert 'amethyst' glass, some fluorescence and lasers

Geometrical and physical optics
12 Dispersive refraction, polarization, etc: rainbows, haloes, 'fire' in gemstones, prism spectrum
13 Scattering: moonstone, red sunset; blue sky, moon, eyes, skin, butterflies, and bird feathers.
14 Interference: oil slick on water, soap bubbles, cracks in minerals, some biological colours.
15 Diffraction; aureole, glory, diffraction gratings, opal, some biological colours, most liquid crystals.

irradiated yellow sapphire (and undoubtedly some similar natural material) are unstable, and fade when exposed to light, while yet others fade even in the dark.

Fluorite sometimes occurs in nature as purple-coloured crystals. Analysis shows no consistently present impurities that might explain the colour. Any colourless fluorite, even synthetically grown crystals of very high purity, can be irradiated to produce this same coloration, attributed to so-called F centres, derived from the German word for colour, *Farbe*. Sylvite also has a purple F centre, halite has a yellow F centre and so on.

2.2.1 F centres

Detailed study has shown that an F centre results when a halide ion such as Cl^- or F^- is missing from its correct position within the crystal and is replaced by a single electron. This is shown in Figure 2.3 for the halide sylvite (KCl). There are always missing ions (either as Schottky or as Frenkel defects) in any crystal, but by themselves these do not produce colour. When energetic radiation, usually from nearby radioactive elements in nature, displaces an electron, this can now be trapped at a Cl^- vacancy to form an F centre.

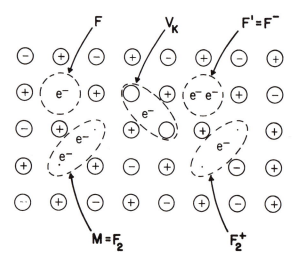

Figure 2.3 Different types of colour centre defects in a halide crystal (schematic).

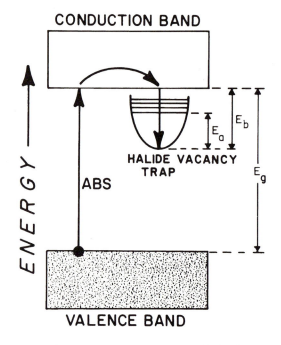

Figure 2.4 Trapping of energy from absorbed light in a halide vacancy trap (F centre) in a halide crystal.

On the basis of 'band formalism'[1] this process has to do with the excitation by irradiation of an electron from the 'valance band' (normally exactly full of electrons) into the 'conduction band' (normally empty). See Figure 2.4. The electron in the conduction band is now attracted to the halide vacancy, because this has an effective positive charge. The vacancy is in fact neutral, but since the structure calls for a negative charge in that position, it acts as if it were positive. Energetically, the vacancy is located within the 'band gap', that is between the valence and the conduction bands, as shown in Figure 2.4.

Once at the vacancy, the electron is 'trapped' by the positive charges of the surrounding cations. There are now additional energy levels available within the trap, as shown in Figure 2.4, and transitions among these levels produce the light absorption leading to the colour of the colour centre. The process of bleaching by heating is then the reverse of the formation process depicted in Figure 2.4, with the electron being excited from the colour centre trap back into the band. Because trapping does not occur at the elevated temperature, the electron can return to its original position in the valance band.

Two energies are of significance in discussing a colour centre. The bleaching energy E_b (see Figure 2.4) determines the temperature at which the colour is lost; since a range of energies is present in thermal vibrations and only energies above E_b will be active, bleaching is both a time-dependent and temperature-dependent process. The light absorption energy E_a, as shown in Figure 2.4, corresponds to the peak of the absorption band. An F centre typically has a broad absorption band, shown in Figure 2.5 for sylvite, where several other colour centres are also shown. There may be one F centre for every 10,000 halide ions in a typical deeply coloured crystal.

Other colour centres can form in halides depending on conditions. During exposure to light, which is absorbed into the F band of Figure 2.5, some of the F centres are converted into F' centres, with the extremely broad band shown in this figure. An F' centre, also called an F^- centre, consists of an F centre at which a second electron has been trapped, as shown in Figure 2.3. The M centre of this figure (also called an F_2 centre for obvious reasons) consists of two adjacent interacting F centres, as shown, absorbing in the infrared region of the spectrum. There is also a group of three adjacent F centres lying in the (111) crystallographic plane, giving the R centre of Figure 2.5. The F_2^+ centre consists of two adjacent halide vacancies with but a single electron trapped between them, as shown in Figure 2.3.

All these centres are 'electron colour centres', because an electron is present in a location where an electron is not normally found. Contrasted to an

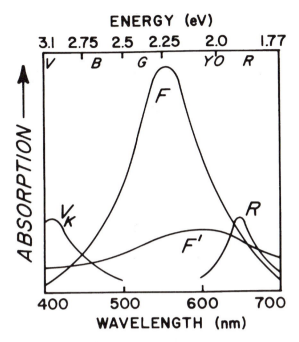

Figure 2.5 Light absorption by colour centre defects in a potassium chloride crystal. After P.W. Levy, 'Color Centers', in Lerner/Trigg, *Encyclopedia of Physics* (Addison Wesley, Reading, MA, 1981), p. 132.

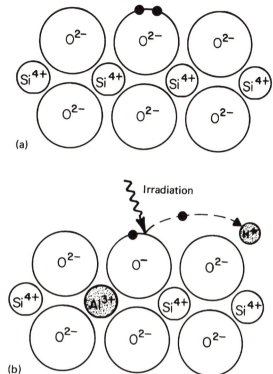

Figure 2.6 Schematic representation of the structure of quartz (A) and the formation by irradiation of a smoky quartz colour centre (B).

electron colour centre is a 'hole colour centre', where an electron is missing from its normal position, producing a light-absorbing centre. An example is the V_k centre of Figures 2.3 and 2.5, where two adjacent halide ions have only one negative charge between them instead of the two normally present. Several dozen other light-absorbing bands are known in the alkali halides, some also involving impurity ions.

This description of F centres also applies in full to fluorites, where the unit + alkali metal anions in Figure 2.4 are replaced by half the number of 2+ calcium ions and the cations are fluoride ions. Light sometimes produces the fading of the purple colour in fluorite, noted by Pearl;[3] this has at times been used to lighten the colour of too dark 'Blue John'.[8] The production of a purple colour in a green fluorite reported by Sweet[10] presumably was caused by colour centre formation from excitation by the ultraviolet present in sunlight.

2.2.2 Smoky quartz and amethyst

When quartz is grown in the laboratory with a small amount of aluminium present, the resulting colourless crystal has Al^{3+} ions substituting for about one out of every 10,000 of the Si^{4+} ions. If this material

is irradiated with x-rays or gamma-rays, the dark brownish-grey to black colour of smoky quartz appears. The colour disappears on heating to around 400°C but can again be recovered on re-irradiation. Natural smoky quartz can be similarly bleached and recoloured, and almost all natural colourless quartz contains enough aluminium to be turned smoky. Figure 2.6a represents the structure of pure quartz, with charges shown for ionic bonding. Two of the spin-paired electrons of the outermost full valence shell of the top central oxygen are shown as black dots. Although irradiation of a pure quartz crystal would eject electrons from some of the oxygens, these would return immediately and the crystal would remain colourless.

Figure 2.6b shows an Al^{3+} replacing one of the Si^{4+} ions, with a proton H^+ (or an Na^+ or a K^+) in an interstitial site to maintain the electrical neutrality of the crystal; this charge-compensating ion is usually located at some distance from the Al^{3+}. If irradiation ejects an electron from an oxygen adjacent to the Al^{3+}, this electron can be trapped by the proton, with the formation of a hydrogen atom:

$$O^{2-} + \text{irradiation} \rightarrow O^- + e^- \qquad (1)$$

or

$$[AlO_4]^{5-} + \text{irradiation} \rightarrow [AlO_4]^{4-} + e^- \qquad (2)$$

and then

$$e^- + H^+ \rightarrow H \qquad (3)$$

The hydrogen atom does not absorb light, but the O^- entity (when in the cluster involving the Al^{3+} with its four surrounding oxygens in the $[AlO_4]^{4-}$ group and containing one unpaired electron as in Equation (2)) does, and it is this group which produces the smoky colour and is the colour centre. Heating releases the electron from the hydrogen atom and reverses first Equation (3) and then Equation (2) or (1). Note that as the colour centre has one electron less than its full complement, it is a hole colour centre.

The bleaching temperature of natural or irradiated smoky quartz varies widely, ranging from as low as 140°C to as high as 380°C.[9] It has been the writer's experience that colour centres that can be bleached by heating to 200°C or less for one or two hours will fade in the light, while those requiring a higher temperature will be stable to light (this can provide a rapid test on a small fragment in place of a tedious fade test; this test should not be used on yellow sapphire, however[35]). Accordingly, some of this smoky quartz could be expected to fade in light. In many instances a greenish-yellow colour (a different colour centre of unknown nature usually co-existing, unseen, together with the smoky colour) is revealed on the transition from smoky to colourless. For a detailed discussion on this topic, together with references, see Nassau.[8,9]

If instead of Al^{3+} there are Fe^{3+} ions present, as in some forms of citrine quartz, then the usual pale yellow colour of Fe^{3+} in a ligand field is seen. On irradiation, the resulting $[FeO_4]^{4-}$ colour centre (produced exactly as in Equation (1) to (3) and Figure 2.6) absorbs light to produce the purple colour of amethyst. The nature of the coloration in natural amethyst was thought to involve such a process, but uncertainty remained until it was duplicated by the growth of synthetic amethyst in the laboratory.[11] Heating bleaches amethyst back to yellow, and re-irradiation returns the colour to both natural and synthetic amethyst, as long as the temperature used in the bleaching is not so high as to produce other irreversible changes in the crystal.

2.2.3 Electron and hole centres

In the most general approach for a colour centre material, consider a crystal containing two precur-

sors A and B, as shown in Figure 2.7a. The hole precursor A has paired electrons, one of which can be ejected by energetic radiation, as shown, while the electron precursor B has the capability of trapping this electron. Neither precursor necessarily absorbs light to produce colour in its original state. After absorption of energy in the irradiation step, the precursor A becomes ionized, with the loss of an electron:

$$A + \text{irradiation} \rightarrow A^+ + e^- \qquad (4)$$

The B precursor also becomes ionized, by trapping this electron

$$B + e^- \rightarrow B^- \qquad (5)$$

The result is shown in Figure 2.7b.

If B^- is the entity which absorbs light, then we have an electron colour centre B^-. Alternatively, if A^+ is the light-absorbing entity, then we have a hole colour centre A^+. Magnetic resonance and related techniques are usually required to establish which of the two centres causes the observed colour. Heating liberates the electron from B^- in Figure 2.7, and the reverse of Equations 5 and 4 restores the crystal to its original colourless state in Figure 2.7a.

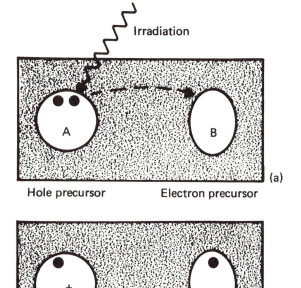

Figure 2.7 The irradiation of hole and electron precursors, (a) to form hole and electron centres (b).

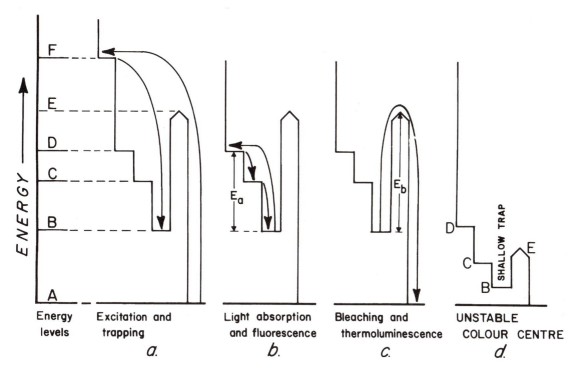

Figure 2.8 The energy levels and the formation of a colour centre (a), light absorption (b), and bleaching (c), all for a stable colour centre. An unstable colour centre (d).

A step energy level diagram can be helpful in thinking about the processes occurring in colour centres. As shown in Figure 2.8a, absorption of energy from the irradiation produces a transition from the ground state A up to level F or into higher energy levels well out in the ultraviolet region of the spectrum. Relaxation with the emission of light or heat results in the trapping of the system into the colour centre trap B. While in the trapped state, the system can absorb light, as in the upward transition B to D as in Figure 2.8b. This absorbs energy E_a as in Figure 2.4 (and could also produce some fluorescence). Finally, heating that supplies enough energy to exceed the B to E barrier, corresponding to energy E_b in Figure 2.4, results in the return to the ground state A as in Figure 2.8c, with the loss of the light-absorbing transition and therefore of the colour. Light emission also sometimes occurs at this stage in the form of thermoluminescence. If E_b is not too large, it is possible for visible light or ultraviolet radiation to empty the trap and bleach the colour; this will only happen if the absorption level F is high enough so that the system cannot be excited to it from the trapped state by the radiation used.

The energy levels of Figure 2.8 refer to the whole system of Figure 2.7. It is interesting to see how the hole and electron centres contribute to individual levels. The ground state A of Figure 2.8 corresponds to the two precursors A and B of Figure 2.7a. The absorption level F of Figure 2.8 is controlled by the properties of the hole centre precursor A in Figure 2.7a. However, the colour-producing absorption levels B, C and D in Figure 2.8 and the energy E_a belong to either the electron or the hole centre (A^+ or B^- in Figure 2.7b), depending on which one produces the colour. The energy involved in the bleaching, that is the depth of the trap E_b, is a property of the electron centre B^- of Figure 2.7b only.

Usually when we speak of the properties of a colour centre, describing its absorption properties and its electron or hole nature, we are referring to either A^+ or B^-, whichever one has the light-absorbing levels. On rare occasions both centres absorb light, as in the hackmanite discussed below. It is, however, not at all rare that a hole colour centre material such as smoky quartz or amethyst may contain several different possible electron centre precursors B, which trap the electron liberated when the hole colour centre A^+ is formed. Specimens from different localities may then have different bleaching temperatures or may even be faded by light, depending on the nature of the B^- and the magnitude of E_b which control this step; while the light-absorbing properties of the colour centre A^+

in these specimens are exactly the same. Since magnetic resonance may be observed either from A$^+$ or B$^-$ or even from both, it is very easy to misidentify the nature of colour centres.

The colour centre illustrated in Figure 2.8 (a–c) has a relatively deep trap. It would accordingly require a fairly high energy to form the colour centre, probably more than the 5 eV of ultraviolet. The colour would be stable to light at room temperature but could fade if heated to several hundred degrees Celsius, as with alkali halide F centres, most smoky quartz, and most amethyst. The colour centre with a very shallow trap illustrated in Figure 2.8d could be formed even by ultraviolet radiation; this material would probably fade in daylight while absorbing part of this light into level C and producing the colour, because the thermal energy of room temperature would assist in surmounting the barrier. An example is the hackmanite discussed below. Essentially all colour centres bleach by about 700 °C, while some are unstable at room temperature in the dark.

2.2.4 Other colour centres

Topaz can contain at least four colour centres: a blue and brown, both stable to light, and two additional unstable browns, having different formation and bleaching rates.[8,12] Several of these can co-exist. Colourless topaz irradiated to become blue must be heated or exposed to light to remove the unstable brown also produced by this process;[12] this same unstable brown overlying a blue has been noted[10] in a natural specimen. The specific natures of the various colour centres in topaz are not known at present, and it is not even clear how many types of blue there are – possibly more than three.

In 1914 the Maxixe mine in Brazil yielded a magnificent deep blue 'Maxixe' beryl, which faded rapidly on exposure to light. More recently a similar but not identical 'Maxixe-type' deep blue beryl has been made by using ultraviolet or other irradiation on certain pale natural beryls. Both these types of blue beryl (not to be confused with the perfectly stable Fe-coloured pale blue aquamarine beryl) fade rapidly because of a shallow trap similar to that shown in Figure 2.8d. The Maxixe colour centre is a hole produced at a nitrate impurity:

$$NO_3^- + irradiation \rightarrow NO_3 + e^- \qquad (6)$$

while the Maxixe-type colour centre involves carbonate:

$$CO_3^{2-} + irradiation \rightarrow CO_3^- + e^- \qquad (7)$$

Both NO_3 and CO_3^- are 23-electron systems with similar symmetry properties; hence the close similarity in colour. For details and references see Nassau.[8]

Quite unusual is the hackmanite variety of sodalite ($Na_4Al_3Si_3O_{12}Cl$), with part of the chlorine replaced by sulphur. Some hackmanite is deep magenta when mined, but fades rapidly on exposure to light. The colour may return on storage in the dark, but always does so on exposure to ultraviolet radiation. Synthetic hackmanite shows exactly the same sequence of colour changes. Thermal excitation at room temperature in the dark may be sufficiently energetic to refill the trap over a period of time, whereas the more energetic light empties it. The mechanism involves an S_2^- hole centre:

$$S_2^{2-} + excitation \rightarrow S_2^- + e^- \qquad (8)$$

This hole centre absorbs at 3.1 eV, but the electron is trapped by a halogen vacancy, forming an F centre absorbing at 2.35 eV; it is the combination of both of these absorption bands that yields the magenta colour.

Further details on colour centres can be found in Nassau,[1,8] Farge and Fontana,[13] Townsend and Kelly,[14] Watts,[15] and Marfunin.[16]

2.3 Other light-induced colour change mechanisms

There are some colour changes which resemble colour centres but are merely valence changes in transition metal ions. One such example is pink kunzite containing both iron and manganese, which acquires a deep green colour on exposure to ultraviolet or other irradiation. In the change the pink of Mn^{3+} is converted to the deep green Mn^{4+} by a coupled oxidation-reduction with Fe^{3+}:

$$Mn^{3+} + Fe^{3+} + irradiation \rightarrow Mn^{4+} + Fe^{2+} \qquad (9)$$

with the reverse process occurring on exposure to light. A similar change occurs in rose quartz, where the colour originates in Ti^{3+}. This colour is bleached on heating to between 200° and 300 °C, or sometimes by bright light, with the formation of colourless Ti^{4+}, and can be recovered with irradiation; it probably also involves Fe. In both these examples the colour is essentially a ligand field colour as in Mechanism 5 of Table 2.2.

Dimorphic transformations can be initiated by light. The darkening of cinnabar comes with the conversion to black metacinnabarite.[17] The red to yellow transformation of realgar, previously thought to be a photo-oxidation to a mixture of orpiment plus As_2O_3, may instead be a transformation to pararealgar[27], at least in some instances.

There are few general principles of utility in thinking about the photochemical changes produced by light or ultraviolet radiation. The first, that only light absorbed can produce changes, is so obvious that it needs no discussion.

The second proposes that one chemical molecule is activated to undergo a reaction for each photon absorbed; this 'law of photo-equivalence' rarely applies, since a chain reaction can produce many molecular reactions from just one photon in some systems, whereas in others most of the absorbed photons may be converted into heat or fluorescence, with but a few producing reactions.

The third principle states that the amount of reaction depends on the number of photons absorbed but not on their energy. However, a minimum energy may be needed to initiate a reaction; in addition, different reactions may be activated at different photon energies. Last, raising the temperature speeds up the photochemical effects, as with any chemical reaction. From the second and third principles, the amount of optical degradation should be linear in time under constant illumination. However, in some systems the decomposition products can protect the remaining material, thus slowing the process; whereas in other systems the decomposition products may speed up the photo-decomposition. There are in fact no generally applicable rules. For general references on photochemistry see Bloomfield and Owsley,[18] Simons,[19] and Wayne;[20] on photochromism see Araujo[21] and Darion and Wiebe.[22]

Light can produce a variety of decompositions. Many silver salts, such as bromyrite and cerargyrite, turn dark because silver is liberated; the initial stages of this process are utilized in photography. The darkening of crocoite appears to come from the liberation of metallic lead and oxygen,[23] possibly:

$$2PbCrO_4 + light \rightarrow Pb(CrO_2)_2 + Pb + 4O \qquad (10)$$

An example of a reaction which may be light accelerated occurs with vivianite, $Fe^{II}(PO_4)_2.8H_2O$. As precipitated in an atmosphere or as the freshly mined mineral, this compound is colourless but soon begins to darken on exposure to light and air. First, it becomes green, then blue, and next a very dark bluish black. In the fresh colourless state all the iron is present as Fe^{2+}. As the oxidation proceeds, the composition can be represented as:

$$Fe^{II}(3-X)Fe_x^{III}(PO_4)_2.(8-x)H_2O.x(OH) \qquad (11)$$

with x increasing from 0 to 3. Because the presence of the OH makes the Fe^{3+} site different from the Fe^{2+} site, Fe^{2+}–Fe^{3+} 'charge transfer' (Mechanism 7 of Table 2.2) can now produce colour, the absorption of light producing the motion of one electron:

$$Fe_A^{2+} + Fe_B^{3+} \rightarrow Fe_A^{3+} + Fe_B^{2+} \qquad (12)$$

where subscript A refers to the normal $2+$ site and B to the normal $3+$ site. For further details see Nassau.[1]

Usually the transformation stops at this state, but with further oxidation the colour sequence can be forced to continue through blue and green to the final yellow ferric phosphate, $Fe^{III}(PO_4)_2(OH)_3$. The colour sequence thus runs from colourless to green, blue, bluish black, blue, green, and yellow.

The deterioration of the colour of organic substances (Mechanism 6 of Table 2.2) by light or the combination of light and atmospheric species is a major problem for museum curators. There can be photo-oxidation, photo-reduction, and even photo-tendering, the actual destruction of a fibre holding a dye.[7] In the field of minerals and gemstones there are only a few instances of such problems.

Amber can be darkened by air-oxidation even in the dark,[24] a process undoubtedly accelerated by light, while dark ivory can be lightened by exposure to sunlight behind glass.[25] Dyeing has been widely used to enhance or alter the colour of gemstone materials, both in the case of porous substances such as chalcedony, turquoise, lapis lazuli, pearl, ivory and jade, as well as in the form of oiling cracks in ruby, emerald and quartz, etc. Some of these processes go back more than 2,000 years.[8] If organic dyes are used, then fading by light is to be expected. Sometimes fading can produce unexpected colour changes, as with an oiled emerald which turned blue because a mixture of blue and yellow dyes had been used, with the latter fading more rapidly than the former. The range of processes used and their identification have been covered by Nassau[8] and King.[4]

Finally, almost any chemical reaction could be speeded up by light. Examples include the light-accelerated sulphide film formation on metallic copper from reaction with H_2S-polluted air;[26] the air oxidation of lower valence transition metal minerals such as marshite, nantockite, rhodochrosite, rhodonite, symplesite; the vivianite transformation discussed above; and the air oxidation of many minerals containing silver sulphide, such as argentite, argyrodite, proustite and pyrargyrite.

2.4 Listings of photosensitive minerals and gems

Tables 2.3, 2.4, and 2.5, respectively, list:

(a) Light-induced colour changes.
(b) Light-induced decompositions,
(c) Light-accelerated surface reactions with air, moisture and/or pollutants.

The items in these tables were compiled from Pearl[2,3], King[4], Sinkankas[7], and Nassau[8], with input

Table 2.3 Light-induced colour changes

Mineral	Note	Effect of Light	Restore	Ref.
Amethyst, *see* quartz				
Anglesite		Brown → C		a
Anhydrite		Blue → C	I	a
Apatite		Mauve or pink → C		a
Barite		C or blue → darker		a, b
Barite		Blue → C		e
Barite		Yellow/brown → green or blue	I	a
Beryl v Maxixe	**	Blue → C or pink	I	c, f
Beryl v morganite	*	Apricot or purplish → pink		b
Beryl v morganite		Pink → paler pink		d
Brazilianite		Green → C	Not I	b, c
Calcite		Fades		a
Celestite	*	Blue → C	I, D	c, f
Chrysoprase, *see* quartz				
Corundum, yellow, fading	#	Yellow → C	I	c
Diamond, chameleon		Yellow → green	D	c
Diamond, chameleon		Red → pink	D	c
Dyed specimens (organic dye)	**	Dye fades	No	c
Fayalite		Green → blue		f
Fluorite		Pink → C		a, c, e, f
Fluorite		Green → purple		a, c, e, f
Fluorite		Blue or purple → C or pink	I	a, c, e
Hackmanite, *see* sodalite				
Haüyn(it)e		Blue → paler		b
Hisingerite		Red → brown		e
Ianthinite		Purple → greenish-yellow		e
Ivory		Lightens		c, f
Kleinite		Yellow → orange	D	e
Jade, mauve (dyed ?)		Fades		
Kunzite, *see* spodumene				
Lepidolite	*	Purple → grey		d
Maxixe beryl. *see* beryl				
Metatyuyamunite		Yellow → green		
Morganite, *see* beryl				
Mosesite		Yellow → green		
Nepheline (nephelite)		Pink → C	D	a
Pabstite		Pink → C		
Phenakite		Red → pink	D	e
Quartz v amethyst		Fades	I	c, f
Quartz v chrysoprase		Fades		a
Quartz v rose	*	Rose → C	?	a, f
Quartz v smoky		Smoky → greenish yellow → C	I	a, b, f
Rose quartz, *see* quartz				
Rutile		Pale → darker	D	a
Sapphire, *see* corundum				
Selenite		Pink fades		a
Smoky quartz, *see* quartz				
Sodalite v hackmanite	#	Red → green, blue, or C	D, I	a, b, e, f
Spodumene (not hiddenite)	#	Green → C	I	b, c, f
Spodumen v kunzite		Pink → C	I	b, c
Topaz	#	Brown → C or blue, rapid	I	c, f
Topaz	*	Brown → C or blue, slow	I	c, f
Topaz		Blue → paler or C		a
Turquoise	*	Fades		a
Vanadinite		Red or yellow → darker		a, e
Zircon		Brown → grey	D	c
Zircon		Brown → blue		e

Abbreviations:	C	Colourless
	*	Can be significant change at times
	**	Drastic change
	#	Rapid change, rarely seen outside of mine
	D	Change reverses in the dark
	I	Change reversed by irradiation
References	a	Pearl (1975)
	b	Sinkankas (1972)
	c	Nassau (1984)
	d	Pearl (1948-1949)
	e	King (1985)
	f	See discussion in text
	g	*CRC Handbook of Chemistry and Physics*

Table 2.4 Light-induced decompostions

Mineral	Note	Effect of Light ·	Ref.
Brom(arg)yrite	**	Darkens, Ag liberated	a, f
Cerargyrite (chlorargyrite)	**	Grey → violet-brown, Ag liberated	a, b, f
Cinnabar	**	Red → black metacinnabarite	a, b, f
Cuprite	*	Darkens, Cu liberated	b
Dyed specimens (organic dye)	*	Dye decomposes and fades	c
Embolite	**	Darkens, Ag liberated	d
Iod(arg)yrite	**	Darkens, Ag liberated	d
Miersite	**	Darkens, Ag liberated	a
Realgar	**	Red → yellow pararealgar	a, b, f

Abbreviations and references see Table 2.3

from discussions with mineral and gem curators and experts and from compilations by Roberts *et al.*,[27] Ford,[28] Palache *et al.*,[29] Frondel[30] and Hey.[31]

Not included in these tables are alterations derived purely from the heating effect of strong light; such as direct sunlight. This can produce the fracturing of sulphur crystals, the volatilization of ammonia salts such as teschemacherite, and the efflorescence of hydrated minerals such as borax, epsomite and melanterite.[6] In this context it is worth mentioning that sunlight can liberate acetic acid from museum cabinets manufactured from oak and a number of other timbers, including birch, resulting in surface attack · on carbonates to produce calclacite[33] and other similar compounds.[34] Also not included is the surface-etching of silver-containing sulphides by very high intensity light;[32] however, this problem is unlikely to occur in the museum environment.

In some of the substances listed in Tables 2.3, 2.4 and 2.5 the photosensitivity is inherent in the material and therefore always present: examples are bromargyrite, Maxixe beryl, and proustite. In other instances the photosensitivity depends on the presence of impurities, or on the specific nature of the defects or impurities, and can vary widely even within one colour of a species. Topaz illustrates the existence in one material of several different colour centres, some fading while others are stable to light, as discussed above. In recent experiments with pink tourmaline irradiated to turn a deep red, the writer noted that light exposure (as well as the equivalent heating to 200°C) faded some of the red but not all of it, indicating more than one colour centre or electron trap. Some coloured calcites fade in the light,[3] presumably from colour centres, while others, namely those containing Fe, Co, or Mn, turn darker, presumably from valence changes.

It should be noted that light-induced alterations may either completely transform a specimen or be self-limiting. The latter occurs, for example, where a surface alteration layer absorbs the photo-active part of the spectrum, thus protecting the interior. Several examples of substances forming opaque tarnish layers are given in Table 2.5.

2.5 Recommendations for protection

In general it is obvious that high-intensity illumination is to be avoided, and that incandescent illumination is slightly preferable to fluorescent illumination. Both of these forms of lighting are much to be preferred to sunlight or daylight, as shown in Table 2.1.

Hinged lids have been used for the more light-sensitive specimens to provide limited public viewing. There is no question that the curator's responsibility is best served by keeping the bulk of such specimens protected from light in the reference section of the collection. This is not necessary with the self-protecting surface-layer materials discussed above, where the layer is already in existence.

The greatest problems arise with those materials where the photosensitivity varies with the origin. Information on specific localities would need to be searched for in each instance; the behaviour may, however, not have been known at the time of the original report. In case of uncertainty the exposure of a small fragment to 200°C for one hour or so has been found by the writer to give a good indication of long-term colour stability to light for colour centres. This test should not be used for yellow sapphire[35].

Where reaction with air, moisture, or pollutants is likely, a number of treatments may be applicable. Coating susceptible specimens with an impervious transparent lacquer or resin may sometimes be effective. Dry or inert atmospheres can be used to protect display specimens.

A list of the most important materials requiring protection is given in Table 2.6. The variability of materials from different sources always presents problems; thus the rare fading of amethyst or smoky quartz is likely to be recognized only after the bulk of the colour has disappeared. The use of irradiation as a means of restoring such specimens[8] might be appropriate in some instances. Clearly much work, both experimental as well as literature-searching, would be required to give more specific recom-

Table 2.5 Light-accelerated surface reaction with air, moisture and/or pollutants

Mineral	Note	Effect of Light	Ref
Acanthite		Alters	a
Aguilarite		Darkens	a
Alabandite		Turns brown	a
Alaskaite		Tarnishes	e
Amber	*	Darkens	c, f
Anapaite		Oxidizes	a
Andorite (sundtite; webnerite)		Yellow	a
Aramayoite			a
Argentite	*	Darkens	a, e
Argyrodite		Turns violet	a
Baumhauerite		Iridescent	a
Berzelianite		Metallic	a
Blockite, *see* penroseite			
Calcite, contg, Fe, Co, or Mn		Darkens	f
Canfieldite		Blue or purple	a
Chalcosite	*	Blue or green	a
Copper		Sulfide film	e
Crocoite		Darkens	a, f
Diaphorite			a
Dietzeite			a
Dufrenoysite		Oxidizes green → brown	a
Dyed specimens (organic dye)	*	Dye oxidizes and fades	c
Eglestonite		Turns brown to black	a
Erythrite			a
Fayalite		Green → brown	a
Fizélyite			a
Freieslebenite			a
Graftonite		Darkens	a
Hessite			a
Hureaulite		Oxidizes	a
Hutchinsonite			a
Ianthinite		Purple → greenish-yellow	a
Jailpaite		Darkens or turns iridescent	d
Kleinite		Darkens	a
Koninckite (strengite)			a
Lengenbachite		Irridescent	a
Lorandite		Yellow	a
Manganosite		Blue	
Marshite		Brownish red	a
Matildite			a
Messelite		Oxidizes?	d
Miargyrite			a
Montroydite		Darkens?	a
Nantokite		Green	a
Naumannite		Brown iridescent	a
Pearceite			a
Penroseite (blockite)		Turns dull	a
Phoenicochroite			a
Polybasite			a
Polydymite		Grey → red or violet	a
Proustite	**	Darkens and silvery	a, b
Purpurite		Darkens	a
Pyrargyrite	**	Darkens	a
Pyrochroite		Turns brown or black	
Pyrostilpnite			a
Ramdohrite			a
Rathite		Iridescent	a
Realgar	**	To As_2S_3 and As_2O_3	a, b, f
Rhodochrosite		Darkens	b
Rhodonite		Darkens	b
Samsonite			a

Sanguinite			d
Sartorite			a
Smithite		Orange	a, e
Sphalerite		Darker and metallic	b
Stephanite		Darkens	a, e
Stibnite		Dark and/or iridescent	a
Strengite, *see* koninckite			a
Stromeyerite		Dark blue	a
Sundite, *see* andorite			a
Sylvanite		Darkens?	a
Symplesite		Oxidizes and darkens	a
Terlinguaite		Yellow → green	a
Trechmannite			a
Vivianite	**	Darkens, can disintegrate	a, c, f
Vrbaite			a
Webnerite, *see* andorite			
Xanthoconite			a

Abbreviations and references: see Table 2.3.

Table 2.6 Recommendations for light protection

Mineral	Effect of Light	See Table
Light protection usually required:		
Beryl v Maxixe	Blue → pink or C	2.3
Brom(arg)yrite	Darkens, Ag liberated	2.4
Cerargyrite (chlorargyrite)	Grey → violet-brown, Ag liberated	2.4
Cinnabar	Red → black metacinnabarite	2.4
Embolite	Darkens, Ag liberated	2.4
Iod(arg)yrite	C → yellow	2.4
Miersite	Darkens, Ag liberated	2.4
Proustite	Darkens and silvery	2.5
Pyrargyrite	Darkens	2.5
Realgar	Red → yellow pararealgar and/or $As_2S_3 + As_2O_3$	2.4, 2.5
Vivianite	Darkens, can disintegrate	2.5
Low light level suggested:		
Amber	Darkens	2.5
Argentite	Darkens	2.5
Beryl v morganite	Apricot or purplish → pink	2.3
Celestite	Blue → C	2.3
Chalcocite	Blue or green	2.5
Cuprite	Darkens, Cu liberated	2.4
Dyed specimens (organic dye)	Dye oxidizes and/or fades	2.3, 2.4, 2.5
Lepidolite	Purple → grey	2.3
Quartz v rose	Rose → C	2.3
Spodumene v kunzite	Pink → C	2.3

mendations. The writer is very much aware of the shortcomings of this survey.

References

Note

Literature coverage is reasonably complete only to the date when this manuscript was submitted.

1 NASSAU, K., *The Physics and Chemistry of Color*, Wiley, New York (1983)

2 PEARL, R.M., *Mineral Collectors Handbook* Mineral Book Co., Colorado Springs, pp. 52-9 (1948-9)

3 PEARL, R.M., *Cleaning and Preserving Minerals*, 4th ed., Earth Science Pub. Co., Colorado Springs, pp. 69-79 (1975)

4 KING, R.J., 'The Care of Minerals, Section 3A, The Curation of Minerals', *J. Russell Soc.*, 1 (3), pp. 94-113 (1985)

5 PARSONS, A.L., 'The Preservation of Mineral Specimens', *Amer. Mineral.*, 7, pp. 59-63 (1922)

6 BANNISTER, F.A., 'The Preservation of Minerals and Meteorites', *Museums J.*, pp. 465-76 (1937)

7 SINKANKAS, J., *Gemstone & Mineral Data Book* Winchester Press, New York, pp. 112-22 (1972)

8 NASSAU, K., *Gemstone Enhancement* Butterworths, London (1984)

9 NASSAU, K. and PRESCOTT, B.E. 'Smoky, Blue, Greenish-Yellow, and other Irradiation-Related Colors in Quartz', *Min. Mag.*, 41, pp. 301–12 (1977)

10 SWEET, J.M., 'Notes on British Barites', *Min. Mag.*, 22, pp. 257–70 (1930)

11 NASSAU, K., *Gems Made By Man* Gemological Institute of America, Santa Monica, CA., (1980)

12 NASSAU, K., 'Altering the Color of Topaz', *Gems and Gemol.*, 21, pp. 26–34 (1985)

13 FARGE, Y. and FONTANA, M.P. *Electronic and Vibrational Properties of Point Defects in Ionic Crystals* North Holland, New York (1979)

14 TOWNSEND, P.D. and KELLY, J.C., *Color Centers and Imperfections in Insulators and Semiconductors* Crane-Russak, New York, (1973)

15 WATTS, R.K., *Point Defects in Crystals* Wiley, New York (1977)

16 MARFUNIN, A.S., *Spectroscopy, Luminescence, and Radiation Centers in Minerals* Springer Verlag, New York, (1979)

17 GETTEN, R.J., FELLER, R.L. and CHASE, W.T. 'Vermilion and Cinnabar', *Studies in Conservation*, 17, pp. 45–69 (1972)

18 BLOOMFIELD, J.J. and OWSLEY, D.C. 'Photochemical Technology', in *Kirk-Othmer Encyclopedia of Chemical Technology*, 3rd ed. Wiley, New York, Vol. 17, pp. 540–59 (1978)

19 SIMONS J.P., *Photochemistry and Spectroscopy*, Wiley, New York (1971)

20 WAYNE, R.P., *Photochemistry* Butterworths, London, (1970)

21 ARAUJO, A.J., 'Photochromic Materials', in *Kirk-Othmer Encyclopedia of Chemical Technology*, 3rd ed. Wiley, New York, Vol. 6, pp. 121–8 (1978)

22 DARION, G.H. and WIEBE, A.F., *Photochromism*, Focal Press, London, (1970)

23 WATSON, V. and CLAY, H.F. 'The Light-Fastness of Lead Chrome Pigments, *J. Oil and Color Chemists Assn.*, 38, pp. 167–77 (1955)

24 BAUER, M., *Precious Stones*, 2 vols Dover, New York, p. 536 (1968)

25 WEBSTER, R. and ANDERSON, B.W., *Gems*, 4th ed. Butterworths, London (1983)

26 GRAEDEL, T.E., FRANEY, J.P. and KAMMLOTT, G.W., 'Ozone- and Photon-Enhanced Atmospheric Sulfidation of Copper', *Science*, 224 [4649], pp. 599–601 (1984)

27 ROBERTS, A.C., ANSELL, A.G. and BONARDI, M., 'Pararealgar, A New Polymorph of A,S from British Columbia', *Can. Mineral.*, 18, pp. 525–27 (1980)

28 FORD, W.E., *Mineralogy of E.S. Dana*, 4th ed. Wiley, New York (1932)

29 PALACHE, C., BERMAN, H. and FRONDEL, C., *The System of Mineralogy of J.D. and E.S. Dana*, 7th ed., Vols. 1 and 2, Wiley, New York (1944, 1956)

30 FRONDEL, C., *The System of Mineralogy of J.D. and E.S. Dana*, 7th ed., Vol 3, Wiley, New York (1962)

31 HEY, M.H., *An Index of Mineral Species and Varieties*, with *Appendices* (British Museum, London (1962, 1963, 1974)

32 STEPHENS, M.M., 'Effect of Light on Polished Surfaces of Silver Minerals', *Am. Mineral.*, 16, pp. 532–49 (1931)

33 VAN TASSEL, R., 'Une efflorescence d'acetatochlorure de calcium sur des roches calcaires dans les collections, *Mus. Belgique Bull.*, 21, [26], pp. 1–11 (1945)

34 FITZHUGH, E.W. and GETTENS, R.J., 'Calclacite and other efflorescent salts on objects stored in wooden museum cases', in *Science and Archaeology*, ed. Brill, R.H., pp. 91–102 (1971)

35 NASSAU, K. and VALENTE, G.K., 'The Seven Types of Yellow Sapphire and their Stability to Light', *Gems and Gemology*, 23, pp. 222–231 (1987)

3

Temperature- and humidity-sensitive mineralogical and petrological specimens

Robert Waller

Mineral specimens have a wide variety of requirements for stability. This is a direct result of the fact that mineral specimens form under a wide variety of conditions.

In nature, diamond will only form at temperatures in excess of 1,000° Celsius and at pressures in excess of 5 gigapascals. Hydrohalite will only form at temperatures lower than zero degrees Celsius, and while it does form at normal atmospheric pressure, the pressure required for its formation is less than one kilopascal.

Acanthite, stable at high temperatures, transforms to argentite before it has cooled to 100° Celsius. It is well known that the mineral ice will melt at temperatures above zero degrees Celsius. At room temperature, lansfordite will dehydrate even at 100% relative humidity. At room temperature, chalcocyanite is hygroscopic at less than 5 per cent relative humidity. Clearly acanthite, ice, lansfordite and chalcocyanite will not survive in collections unless they are given special care.

Probably more than 10% of the approximately 3,000 known mineral species are susceptible to alteration or complete destruction when exposed to commonly encountered environmental conditions. Fine display specimens are destroyed and important reference specimens are altered simply because of exposure to an inappropriate level of temperature or relative humidity. In many cases the climatic limits required for specimen stability can be quantified precisely; hence little excuse exists for subjecting a specimen to conditions outside these limits. In other cases climatic requirements are difficult to quantify. Fortunately, however, awareness of the potential problems and a measure of care and common sense in handling such specimens can all but guarantee their preservation.

In all cases knowledge, either of specific stability limits or of tendencies to change in response to changes in environment, is a prerequisite to providing proper care for collections. This chapter provides this knowledge and describes the methods available for the prevention of temperature- and humidity-related damage to specimens.

3.1 Temperature

3.1.1 Polymorphic phase transitions

The temperature of formation for many mineral species is in excess of 300°C. This is somewhat higher than the temperatures normally encountered in collection storage areas, so it is not surprising that many minerals are not in a stable form at room temperature. Two of the high temperature polymorphs of quartz, tridymite and cristobalite, are examples. Below 573°C low quartz is the only stable form of silicon dioxide, and hence, from a thermodynamic standpoint, both tridymite and cristobalite should convert to low quartz when they are stored at room temperature. Both these minerals do persist, however, since the rate of transition at room temperature is so slow that any change is imperceptible; hence, they are said to be in a metastable state.

In contrast, high quartz, another high temperature polymorph of quartz, is never encountered in specimens that are at room temperature, since the transition to low quartz is easy and proceeds rapidly.

Generally mineral species which are unstable with respect to another polymorphic phase at room temperature will either have converted to that phase by the time they are collected, or will be so slow in their transformation that they do not pose a preservation problem.

3.1.2 Volatilization

Volatilization is the loss of a chemical substance through conversion to a vapour. In mineral specimens this loss may occur through evaporation, sublimation, or dissociation. The basic driving force for volatilization is a difference between the vapour pressure of gases derived from a mineral and the partial pressure of those gases in the air surrounding the specimen. Volatilization has been included as a temperature-related form of deterioration, since the vapour pressure of a substance is temperature-dependent.

Parsons[1] listed sal-ammoniac and teschemacherite as minerals that are subject to damage through volatilization. In the case of sal-ammoniac, NH_4Cl, the vapour pressure at normal room temperature is very low, of the order of 10^{-3} Pa (calculated from data in Dean).[2] Sublimation would be expected to be very slow indeed in the case of this mineral. Teschemacherite, NH_4HCO_3, on the other hand, exerts a relatively high vapour pressure at room temperature, 7.9 kPa at 25.4°C.[3] For comparison this is more than twice the vapour pressure of pure water at the same temperature.

Several other minerals have vapour pressures intermediate to those above. The vapour pressure of mercury is 0.24 Pa at 25°C,[4] and the vapour pressure of rhombic sulphur is 0.17 Pa at 25°C.[5] While these are not high enough to cause dramatic losses from mineral specimens, they are high enough that some loss from specimens could be expected. In the case of mercury, specimens should be stored in sealed containers to prevent the escape of toxic mercury vapours for reasons of health as well as for preservation. It is not yet certain whether or not sulphur crystals are significantly damaged by sublimation. While the magnitude of the vapour pressure suggests that damage would occur, it is possible that adsorbed contaminants are sufficient to prevent significant sublimation.

The dissociation of hydrated minerals to give water vapour and a solid, lower hydrate is the most common type of dissociation reaction. This reaction will be dealt with separately in Section 3.2. The volatilization of teschemacherite is in fact a dissociation reaction, since the gases evolved are NH_3, H_2O and CO_2 rather than gaseous molecular NH_4HCO_3.

That dissociation is involved in volatilization of CO_2 from carbonates is clearer in cases where at least one of the products of dissociation is solid. Figure 3.1 is a part of the phase diagram for the system $NaOH-CO_2-H_2O$. It has been drawn to relate percentage relative humidity to the partial pressure of carbon dioxide. This diagram shows that at the levels of carbon dioxide partial pressure normally encountered trona is the most stable phase at all except very low levels of RH. At RH levels above about seven per cent all phases will tend to convert to trona. For nahcolite this reaction involved both hydration and dissociation, and proceeds according to the reaction:

$$3\ NaHCO_3 + H_2O \rightarrow Na_2CO_3 \cdot NaHCO_3 \cdot 2H_2O + CO_2$$

The enthalpy of reaction is 51.8 kJ mol^{-1}; hence the reaction is endothermic and is favoured at high temperatures. Assuming $\Delta_r H^\ominus$ is constant over the temperature range of interest, and using the equilibrium constant for 25°C calculated from free energy data, one can use an integrated form of the van't Hoff isochore:

$$\ln(K_{T2}) = \ln(K_{T^*}) - (\Delta_r H^\ominus/R) \times (1/T2 - 1/T^*)$$

where: K_{T2} = equilibrium constant at T2
  K_{T^*} = equilibrium constant at T*
  $\Delta_r H^\ominus$ = enthalpy of reaction
  R = molar gas constant
  T^* = 298.15 K
  $T2$ = temperature sought

This can be used to calculate the temperature at which nahcolite will be more stable than trona, assuming RH = 50% and P_{CO2} = 22.5 Pa, this temperature would be -10°C. Consequently, a specimen of nahcolite would be stable if it were stored in a freezer at less than -10°C at 50% RH.

Natrite may be maintained by storage at less than 6% RH. However, under these conditions associated minerals may be altered. The hydrated sodium carbonate minerals could be safely stored at relative humidity levels at which they are stable if a CO_2 sorbent more efficient than the respective sodium carbonate hydrate was enclosed with the specimen in a tightly sealed container and replaced as necessary.

Specimens enclosed in hermetically sealed containers will react with the atmosphere within that container only until equilibrium concentrations of water vapour and carbon dioxide have been established. Unless the container is very large in comparison with the amount of reactive material enclosed in it, then the amount of specimen that must react to establish this equilibrium would probably be negligible. The arguments between storage at enforced conditions and storage at specimen equilibrium conditions will be discussed in a later section.

Numerous other carbonate minerals are also subject to dissociation or association reactions. Where thermodynamic data are available, diagrams similar to Figure 3.1 can be drawn to determine which phases might be subject to these reactions. The method of creating such diagrams is described in many texts on phase equilibria; one particularly useful text is that by Garrels and Christ.[6] It should

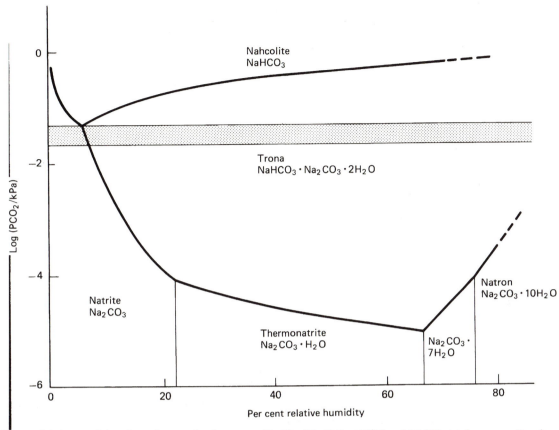

Figure 3.1 A part of the phase diagram for the system Na_2CO_3-CO_2-H_2O at $25°C$ and 0.1 MPa total pressure. Based on free energy data.[6] Normal PCO_2 levels[7] are indicated by the shaded area. PCO_2 levels may be two to three times this high in some buildings, particularly those with high people loads and low ventilation rates.

be remembered that these diagrams indicate thermodynamic stability only. It is possible that some phases will persist in a metastable state even though they are maintained at conditions under which they are not stable. If the mineral species concerned has been known for some time, then attempts to identify it and/or its dissociation–reaction products in old museum specimens that were originally identified as containing that species may indicate whether or not the mineral is capable of persisting metastably. If the mineral species concerned is a new species, then it is best to treat at least part of the type material as though it were unstable.

Generally, maintaining volatile specimens at low temperatures will reduce rates of volatilization. In fact, as illustrated in the case of nahcolite, if the product of dissociation is normally present in the atmosphere, then volatilization can be eliminated by storing specimens at a temperature at which the partial pressure of the dissociated gas is lower than the ambient partial pressure of that gas. Most

frequently, however, the most practical means of preservation for specimens subject to volatilization is storage in hermetically sealed containers.

3.1.3 Dissociation of hydrates

As mentioned in the previous section, the special case of volatilization where hydrates dissociate to give water vapour and a lower hydrate will be discussed in detail in Section 3.2, since the survival of these species primarily depends on what relative humidity levels they are exposed to. Nevertheless these species will dissociate at any relative humidity if they are exposed to too high a temperature. For some species this temperature is low enough that they must be considered temperature-sensitive. These species will be discussed here.

The high temperature limit to the stability field of most hydrated mineral species is defined by the intersection of the mineral-lower hydrate phase boundary with the mineral-solution phase boundary,

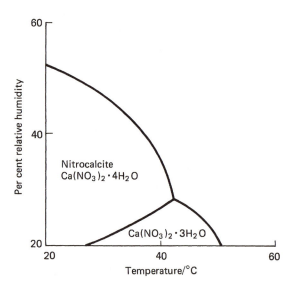

Figure 3.2 A part of the relative humidity–temperature phase diagram for the system $Ca(NO_3)_2$–H_2O. Based on data of Ewing.[8]

Table 3.1 Temperatures at which selected mineral species convert to form a saturated solution of the next lower hydrate

Species	Formula	Temperature	Reference
Antarcticite	$CaCl_2 \cdot 6H_2O$	29.9	9
Bianchite	$ZnSO_4 \cdot 6H_2O$	48.8	10
Bieberite	$CoSO_4 \cdot 7H_2O$	43.3	9
Epsomite	$MgSO_4 \cdot 7H_2O$	48	10
Goslarite	$ZnSO_4 \cdot 7H_2O$	37.9	10
Hydrohalite	$NaCl \cdot 2H_2O$	0.1	10
Lansfordite	$MgCO_3 \cdot 5H_2O$	~10	11
Mallardite	$MnSO_4 \cdot 7H_2O$	9	10
Mirabilite	$Na_2SO_4 \cdot 10H_2O$	32.4	10
Morenosite	$NiSO_4 \cdot 7H_2O$	31.5	10
Natron	$NaCO_3 \cdot 10H_2O$	32.0	10
Nitrocalcite	$CaNO_3 \cdot 4H_2O$	42.7	9

For species in which solid solution is possible the actual transition temperatures will vary from those given here.

as shown in Figure 3.2. Exposure of a mineral to temperatures beyond this point will result in dissociation, regardless of what RH the sample is exposed to. The lower hydrate will dissolve in the water of crystallization released in the dissociation to an extent that will depend on the solubility of the mineral.

In the case of lansfordite, $MgCO_3 \cdot 5H_2O$, dissociating to give nesquehonite, $MgCO_3 \cdot 3H_2O$, the very slight solubility of $MgCO_3$ will result in very little of the mineral going into solution. The reaction will appear to involve only solid and vapour phases. In contrast, in the case of nitrocalcite, $Ca(NO_3)_2 \cdot 4H_2O$, dissociating to give the trihydrate the solubility at the transition temperature is sufficient for all the mineral to go into solution, and, hence, the mineral will melt completely. Most mineral species behave in a manner intermediate to these extremes, and form a mixture of the saturated solution and the lower hydrate in solid form.

The dissociation temperatures for a number of hydrated mineral species that dissociate below 50°C are given in Table 3.1.

3.1.4 Fluid inclusion decrepitation

Many minerals contain cavities that are partly or completely filled with fluid, and are called fluid inclusions. In some cases the fluid is under sufficient pressure to explode spontaneously, as in the case of 'popping rocks'. These vesicular, glassy lava pebbles, dredged from the Mid-Atlantic Rift, jump and explode loudly within days of being dredged from the sea floor (P ~30 MPa) and placed on a ship's deck (P ~0.1 MPa).[7]

In a case related by Roedder[8] a quartz crystal in a museum display case exploded spontaneously. No mention was made of the manner of lighting used for the display case, but it is probable that warming of the crystal by the source of illumination, particularly if it was sunlight, resulted in an inclusion pressure sufficient to decrepitate the crystal. There have been numerous reports of specimens decrepitating violently under intense lighting. The basic cause of this is that the coefficient of thermal expansion of the fluid and/or gas contained in the inclusion is higher than that of the mineral.

It is possible, from studies of fluid inclusions in a particular specimen, to determine dependence of inclusion pressure on specimen temperature, but this information is insufficient for determining the temperature at which the specimen will decrepitate. In determining whether or not an inclusion will decrepitate at a particular temperature, the mechanical strength of the specimen as a container for the inclusion is equally as important as the pressure of the inclusion. Factors that will tend to lower the temperature of decrepitation for an inclusion include:[8]

- Increased inclusion size.
- Increased irregularity and jaggedness of inclusions
- Increased abundance of inclusions.
- Arrangement of groups of inclusions.
- Lower toughness and increased brittleness of the mineral
- Decreased depth of inclusions within specimens.

The significance of the last factor is illustrated in Figure 3.3, where an inclusion near the surface of a

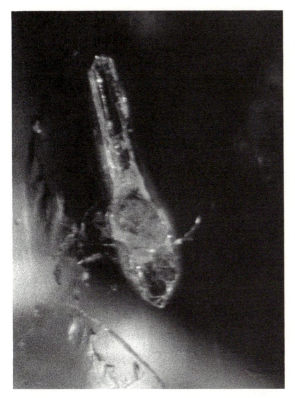 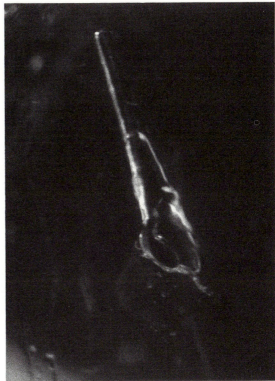

Figure 3.3 Two fluid inclusion cavities in a quartz crystal. CMN #51364. Length of each inclusion is ~1 mm. (a) An inclusion which was very close to the surface has decrepitated. (b) An inclusion, ~1 mm below the surface, has survived. Note the gas bubble in the intact inclusion.

quartz crystal has decrepitated while a nearly identical inclusion at slightly greater depth has survived.

As a consequence of the large number of factors affecting the decrepitation temperature in specimens, it is not possible to set any particular temperature as being the maximum safe temperature for any given specimen. By taking care not to expose specimens to heat, however, through careful planning of exhibit lighting, moderate lighting for specimen photography, and care in specimen transportation in hot weather, one can minimize the chances of specimen damage through fluid inclusion decrepitation.

In addition to spontaneous decrepitation and decrepitation at elevated temperatures, crystals are occasionally subject to fracturing through the freezing of water contained in fluid inclusions. An account of this appeared as early as 1820 when Dwight[9] described quartz crystals from the Kaatskill (Catskill) Mountains, New York. These crystals fractured when they were exposed to a temperature less than $-22°C$. This sort of damage would only result from specimens being stored in unheated rooms or transported by unheated carriers in regions

and during seasons where very low temperatures are expected. Preservation of these specimens simply depends on avoiding exposure of susceptible specimens to these conditions.

3.1.5 Thermal shock

Damage to mineral specimens directly attributed to the rate of change of temperature occurs in the form of fracturing from thermal shock. Crystals will have one, two or three coefficients of linear thermal expansion if they are (1) isometric, (2) trigonal, hexagonal or tetragonal, or (3) orthorhombic, monoclinic or triclinic respectively.[10] Most commonly all coefficients are positive, meaning that the mineral will expand in all directions on heating. Heating a mineral with all coefficients of expansion being positive will result in compressional stress in the outer, heated areas, and tensile stress in the inner, relatively cool areas[11] (Figure 3.4). One type of failure commonly observed in this situation is spalling – a fracturing along a plane at an angle of about 45° to the surface.[12] Commonly this will occur across a corner of a crystal.

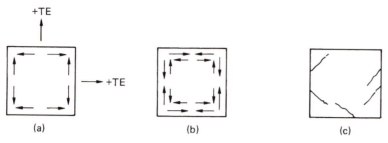

Figure 3.4 (a) Tendency for expansion, ← →, in a crystal having positive coefficients of thermal expansion (+TE) and being subjected to an increase in ambient temperature.
(b) Compressional, → ←, and tensional, ← →, stresses resulting from outer portions of a crystal expanding.
(c) Crack patterns to be expected from brittle failure of a crystal exposed to stress as in (b).

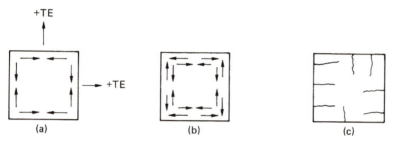

Figure 3.5 (a) Tendency for contraction, → ←, in a crystal having positive coefficients of thermal expansion (+TE) and being subjected to a decrease in ambient temperature.
(b) Compressional, → ←, and tensional, ← →, stresses resulting from outer portions of a crystal contracting.
(c) Crack patterns to be expected from brittle failure of a crystal exposed to stress as in (b).

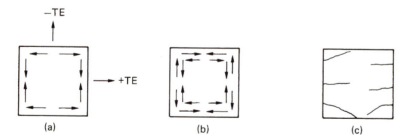

Figure 3.6 (a) Tendency for expansion, ← →, and contraction, → ←, in a crystal having both positive (+TE) and negative (–TE) coefficients of thermal expansion and being subjected to an increase in ambient temperature.
(b) Compressional, → ←, and tensional, ← →, stresses resulting from outer portions of a crystal expanding and contracting.
(c) Crack patterns to be expected from brittle failure of a crystal exposed to stress as in (b).

Spalling occurs when sulphur crystals are held tightly in the hand. The crystals are so heat-sensitive that body warmth transferred to a large crystal will cause it to spall audibly. Cracking through the centre of crystals may also occur as a result of the tensile stress developed in response to compression of the outer, heated areas.

Rapidly cooling a crystal that has all positive linear coefficients of thermal expansion will induce the opposite stresses, i.e. tensile stress in the outer, cooler areas, and compressional stress in the inner, relatively warm areas. The mode of failure under these stresses will be fracturing roughly perpendicular to surfaces (Figure 3.5).

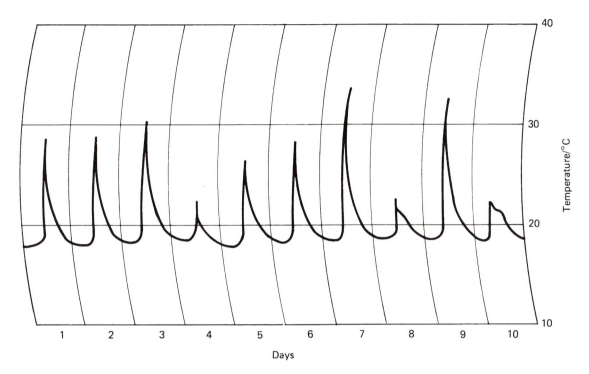

Figure 3.7 Thermograph made in a glass top exhibition cabinet exposed to direct sunlight for about 3 hours per day. Courtesy of Paul Hicks.

Certain minerals, most notably carbonates with the calcite structure, have both positive and negative coefficients of thermal expansion. In the case of calcite the coefficient is positive in directions perpendicular to the C-axis and negative parallel to the C-axis. When a calcite crystal is heated, stress in outer areas may be either compressional or tensile, depending on the orientation of the faces concerned (Figure 3.6).

The fracture patterns described above pertain to brittle minerals without cleavage. Where cleavage is present, fracturing will probably occur along cleavage planes. In addition, rapid crack development may result in fractures being propagated through areas under little or no stress and may lead to complete splitting of crystals.

Characteristics of a mineral specimen that would contribute to sensitivity to thermal shock include:

● High coefficients of thermal expansion.
● Low thermal conductivity.
● High brittleness.
● Low toughness
● Easy cleavage.
● Presence of incipient fractures or cleavages.
● Presence of fluid inclusions, especially if arranged in planes.

● Large, unbroken crystals.
● Low albedo (when radiant heating is the problem).

In addition environmental factors that would contribute to thermal shock include:

● Large temperature change.
● High rate of heat exchange with the specimen.

The magnitude of stress required to produce a fracture may depend on the type of flaws that are present to ease the initiation of cracks. Because of this, it is difficult to make precise statements regarding the severity of thermal shock that any particular specimen may be able to withstand. Generally, large, clear crystals with easy cleavage are the most likely to be affected. Crystals that are warm to touch, indicating low thermal conductivity, are more likely to be thermal sensitive than similar crystals that are cold to touch.

3.1.6 Effect on relative humidity

Changes of temperature will cause changes in relative humidity within containers if humidity buffering materials within the containers are unable to compensate sufficiently rapidly and/or to the extent necessary to mitigate the change. While problems with specimen exposure to inappropriate levels of relative humidity are the topic of Section 3.2, it is important to remember that attempts to maintain specimens at appropriate levels of relative humidity may be ineffective if ambient temperatures fluctuate widely.

3.1.7 Prevention of temperature-related preservation problems

Prevention of temperature-related damage to most specimens commonly requires only awareness of the possible problems and a measure of common sense in handling susceptible specimens.

Extra care is required when exhibits are being designed, to ensure that lighting will not cause excessive heating. Direct sunlight is particularly dangerous, with respect to both high absolute temperatures and rate of temperature increase. This is shown dramatically in Figure 3.7, where the recorded temperature in a horizontal glass-topped showcase exposed to direct sunlight is shown for a one-week period. Temperatures in excess of 33°C were recorded, with rates of temperature change being as high as 8°C/hour.

Lighting for photography is another potential cause of both high rates of temperature increase and high absolute temperatures. Such lighting should be kept to a minimum. The use of flash lighting is recommended for particularly sensitive materials.[13]

In transportation, particularly in harsh climates, specimens are often exposed to extremes in temperature (Figure 3.8). Sensitive specimens should only be shipped by a carrier able to provide temperature control. If specimens that can tolerate extreme ambient temperatures, but not sudden changes of temperature, must be transported by common carrier, then they should be packed in well-insulated crates, and, before unpacking, they should be allowed to sit in their crates overnight to allow for a gradual attainment of thermal equilibrium.

Preparation of specimens requiring immersion in a fluid, such as washing, selective chemical dissolution treatments, or rinsing, may lead to thermal shock if the specimens and all liquids employed are not maintained at the same temperature. Temperature differences between a specimen and a surrounding liquid are more destructive than equivalent differences between a specimen and air, owing to a greater rate of heat transfer. Particular care is required when ultrasonic baths are employed, since they increase the temperature of liquids during a treatment. Finally, with crystals that are highly susceptible to thermal shock, the rate of drying may have to be regulated to avoid excessive evaporative cooling. Certainly such specimens should not be placed in an airstream to speed drying.

Refrigeration is required for safe storage of a relatively small number of mineral species. Many museums with only a few such specimens may find it advantageous to store these in a refrigerator or freezer employed primarily for biological specimens, though this carries the inconvenience of having part of the collection physically separated from the

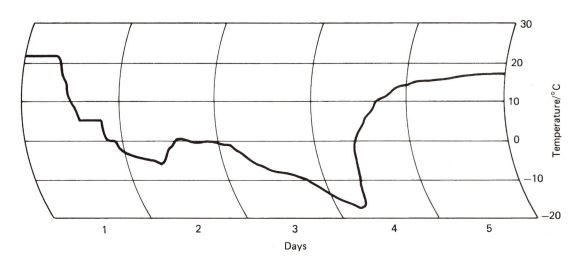

Figure 3.8 Thermograph made during shipment of a 65 × 45 × 30 cm crate of mineral specimens sent by truck from Ottawa to Fredericton in February 1985. Extremes in both absolute temperature and rate of change of temperature are common when specimens are shipped in a harsh climate.

remainder. This type of refrigerator or freezer should be equipped with an auxiliary power supply and a high-temperature alarm, making it well suited for preservation purposes.

3.2 Relative Humidity

In an average collection environment inappropriate levels of relative humidity (RH) cause more deterioration to mineral specimens than any other environmental factor. There are at least four distinct ways in which minerals are altered by improper levels of relative humidity: corrosion, phase transitions, loss of zeolitic water and mechanical failure resulting from hygrostatic stress.

3.2.1 Corrosion

The word 'corrosion' is used here to include any transformation a mineral may undergo by reaction with one or more atmospheric gases other than water vapour. Water vapour may be instrumental or even essential in the progress of these reactions, but reactions which involve water vapour alone are not included in this definition, so that they can be discussed under more appropriate headings.

The greatest problems in mineral corrosion, in terms of numbers of specimens affected, are oxidation reactions. The oxidation of pyrite, commonly termed 'pyrite disease', is the most familiar of these reactions. Howie[14] has shown that oxidation rates in oxidation-susceptible specimens increase rapidly with increasing relative humidity. While some oxidation occurs at all RH levels, the rate of oxidation increases greatly in the area of 60% RH. These results have clearly established the necessity of maintaining relative humidity in the vicinity of susceptible specimens below 60%, and preferably nearer 30%. For a more detailed treatment of pyrite oxidation, see Chapter 6.

The effect of relative humidity and temperature on the oxidation of pyrrhotite has been studied in depth by Steger.[15] It was found that, at 50°C, rapid oxidation occurred at RH 50% as a result of a $Fe(OH)(SO_4) \cdot xH_2O$ product layer undergoing hydrolysis according to the reaction:

$$Fe(OH)(SO_4) + H_2O \rightarrow FeO(OH) + H_2SO_4$$

The mechanism of formation of $FeO(OH)$ was independent of temperature over the range 28° to 50°C. This temperature range is sufficiently close to room temperature for it to be assumed that at room temperature there exists a critical relative humidity for the disruption of the basic sulphate coating and the consequent rapid oxidation of pyrrhotite.

While it is clear that low levels of relative humidity are essential for the preservation of oxidation-susceptible sulphides in specimens, it should be remembered that they are not always sufficient. Some form of treatment is required for specimens which have started oxidizing if they are to be kept in anything other than completely dry or oxygen-free atmospheres. For a more detailed treatment of mineral oxidation problems see Chapters 5 and 6.

The effect of high relative humidities on iron, particularly for iron in contact with water-soluble impurities, is well known. Terrestrial native iron does occur, but most native iron in geological collections will be in meteorites. The conservation of these is dealt with in Chapter 7.

Regarding corrosion of other materials, Padfield, Erhardt and Hopwood[16] have suggested that films of solutions formed by the interaction of pollutants with atmospheric moisture may be at least as important in promoting deterioration as the actual concentration of pollutant alone. Rice *et al.*[17] have presented evidence that these films are not continuous, but include local, deeper accumulations of solution, and have proposed that these local accumulations of solution are the cause of significant corrosion. The formation of such solution films can be minimized by maintaining low levels of relative humidity, in addition to reducing concentrations of pollutants.

In summary, many types of corrosion are known to proceed more rapidly at high levels of relative humidity. Consideration of corrosion problems would lead to a preference for maintaining low levels of relative humidity in mineral collections.

3.2.2 Humidity-related phase transitions

A pressure-temperature (P-T) diagram for the system Na_2SO_4-H_2O is given in Figure 3.9. The phase boundaries for the metastable 7-hydrate have been omitted for clarity. Each area in this diagram has been labelled with the solid or liquid phase which coexists with water vapour under those P-T conditions. At P-T conditions, defined by the lines dividing these areas, three phases are able to co-exist. On line 1 thenardite, mirabilite and vapour co-exist; on line 2 mirabilite, saturated solution, and vapour co-exist; and on line 3 thenardite, saturated solution and vapour co-exist.

Diagrams of this type can easily be converted to relate percentage relative humidity to temperature (%RH-T), as shown in Figure 3.10. This diagram contains the same information as Figure 3.9, but displays more clearly the effects various relative humidity levels will have on specimens of mirabilite or thenardite.

Consider, for example, a mirabilite specimen in a container at the conditions marked by an 'X', that is

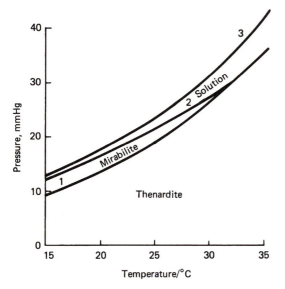

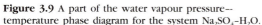

Figure 3.9 A part of the water vapour pressure--temperature phase diagram for the system Na_2SO_4–H_2O.

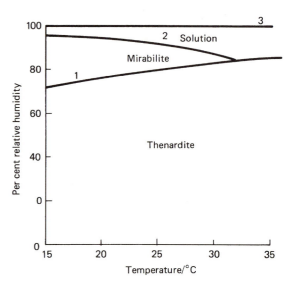

Figure 3.10 A part of the relative humidity-temperature phase diagram for the system Na_2SO_4–H_2O.

at 85% RH and 20°C. The container could be a jar, box, storage cabinet or exhibition case, all of which are imperfect containers and will allow passage of water vapour. If water vapour penetrates the container, the RH will rise until line 2 is reached at 93% RH. At this point the saturated solution becomes a stable phase, and further addition of water vapour will result in the formation of solution at the expense of mirabilite and water vapour. This spontaneous dissolution by atmospheric moisture is known as deliquescence, and mineral species prone to reacting in this way are described as being deliquescent. The term deliquescent is relative. Halite, which deliquesces at 75% RH, would not be considered a deliquescent mineral by observers in very arid areas, but it would be considered deliquescent by observers in tropical areas.

Returning to the specimen at 85% RH and 20°C (Figure 3.10), if water vapour escapes from the container the RH will fall until line 1 is reached at 75% RH. At this point thenardite becomes a stable phase, and further escape of water vapour will result in the formation of thenardite and water vapour at the expense of mirabilite. This spontaneous loss of water of crystallization is known as efflorescence, and mineral species prone to reacting in this way are described as being efflorescent. The term efflorescent is relative. Chalcanthite, which effloresces at 33% RH, would not be considered efflorescent by observers in tropical areas, but it would be considered efflorescent by observers in arid areas.

Similarly, a specimen of thenardite exposed to RH-T conditions above line 1 (Figure 3.10) will undergo hydration to mirabilite.

3.2.3 Deliquescence

Deliquescence can occur simply, can lend to decomposition or can be accompanied by other reactions such as hydrolysis or oxidation.

Simple deliquescence can be either intermittent or continuous. Continuous deliquescence will occur when the relative humidity in the area of the specimen is constantly above the stability limit of the mineral species present. When this happens, the mineral will exist only in solution, and will appear only as a puddle of liquid or, if the specimen was on an absorbent surface, as a stain in that surface. Deliquescence will be intermittent if the relative humidity in the area of the specimen is only above the stability limit of the specimen part of the time. The extent to which a specimen will deliquesce during a period of high relative humidity will depend on the RH level, the duration of exposure, the rate of air flow, the total amount of moisture available, and on characteristics of both the species and the specimen in question. A general rounding or flattening of specimens, the result of flow of the solution phase, is the usual consequence of intermittent deliquescence. In extreme cases specimens can be reduced to mere crusts in the bottom of their trays.

Figure 3.11 Hanksite crystals coated with a mixture of thenardite and other minerals precipitated from the solution formed during a period of deliquescence. CMN #34749, width of specimen as viewed is 10 cm. (Photograph by G. Runnels.)

Figure 3.12 Tincalconite after borax, CMN #38207; width of specimen as viewed is 13 cm. These powdery pseudomorphs are much more common in mineral collections than are the transparent borax crystals from which they form, since the transition proceeds rapidly at relative humidity levels lower than 50 per cent. (Photograph by G. Runnels.)

Deliquescence can result in decomposition of a mineral species if the mineral contains more than a single cation and/or anion. An example of this is buetschliite, $K_2Ca(CO_3)_2$. When exposed to moist air K_2CO_3 is leached from crystals of this mineral, leaving a powdery pseudomorph of $CaCO_3$.[18]

Hanksite, $KNa_{22}(SO_4)_9(CO_3)_2Cl$, is another example of a mineral that undergoes decomposition during deliquescence. At normal room temperatures hanksite is not the least soluble phase in the solution formed as hanksite dissolves. Figure 3.11 shows the pulverulent coating that forms on hanksite crystals following a period of deliquescence and subsequent drying. This coating has the same bulk composition as hanksite, but is composed of thenardite, Na_2SO_4, plus at least two other mineral phases. Most salt systems have been studied in sufficient detail for reference to the chemical literature to reveal whether or not a salt will dissolve congruently at room temperature, and hence whether or not it will undergo decomposition during deliquescence.

Certain minerals undergo hydrolysis when they deliquesce. Plate 3.1 shows the effect of hydrolysis on coquimbite, $Fe_2(SO_4)_3 \cdot 9H_2O$, crystals. Formation of the iron oxyhydroxide crust on these crystals is accompanied by formation of sulphuric acid that migrates down through the specimen.

Oxidation, for example $Fe^{2+} \rightarrow Fe^{3+}$, can also occur while minerals are in solution. Melanterite, $Fe^{2+}SO_4 \cdot 7H_2O$, for example, has been observed to partially transform to bilinite, $Fe^{2+}Fe^{3+}_2(SO_4)_4 \cdot 22H_2O$, during storage in a slightly damp state (Plate 3.2).

3.2.4 Efflorescence

The effect of efflorescence on specimens varies with the nature of the species. It can be anything from the development of one or just a few shrinkage fractures to the complete decrepitation of a crystal into a formless mass of microcrystalline powder.

Autunite is an example of a species that can undergo efflorescence without decrepitating structural changes. The meta-autunite crystal shown in Plate 3.3 formed as a result of the dehydration of an autunite crystal. The crystal has not decrepitated but has simply opened along the {001} cleavage plane in response to the shrinkage that occurs perpendicular to that plane during dehydration.

More noticeable damage occurs when borax, $Na_2B_4O_5(OH)_4 \cdot 8H_2O$, dehydrates to tincalconite, $Na_2B_4O_5(OH)_4 \cdot 3H_2O$. Borax crystals are often clear and transparent when fresh, but become chalky and friable after dehydration (Figure 3.12). The bonattite, $CuSO_4 \cdot 3H_2O$, that results from the efflorescence of chalcanthite, $CuSO_4 \cdot 5H_2O$, is so loosely aggregated that it is incapable of retaining even the original shape of the chalcanthite, and will fall into a formless mass of powder as the transformation approaches completion. The earliest stage of physical disintegration of the bonattite that results from the efflorescence of chalcanthite is apparent in the development of shrinkage fractures, as shown in Plate 3.4.

3.2.5 Hydration

Many well-formed crystals will persist indefinitely in a metastable state at relative humidity levels that

Table 3.2 Minerals subject to humidity-related phase transitions

Species	Formula[1]	Reaction[2]	%RH[3]	T[4]	Ref[5]
Acetamide	CH_3CONH_2	deliquesce			1
Aluminite	$Al_2(SO_4)(OH)_4 \cdot 7H_2O$	$-nH_2O$			2
Alunogen	$Al_2(SO_4)_3 \cdot 17H_2O$	deliquesce	88	20	2
		$-nH_2O$			3
Antarcticite	$CaCl_2 \cdot 6H_2O$	deliquesce	33	RT	c4
		$-2H_2O \rightarrow 4$	21	RT	c5
Anthonyite	$Cu(OH,Cl)_2 \cdot 3H_2O$	$-nH_2O$			6
Arsenuranospathite	$HAl(UO_2)_4(AsO_4)_4 \cdot 40H_2O$	$-nH_2O$			7
Autunite	$Ca(UO_2)_2(PO_4)_2 \cdot 10-12H_2O$	$-nH_2O \rightarrow$ meta	~40	25	i8
Bandylite	$CuB(OH)_4Cl$	deliquesce			9
Barentsite	$Na_7AlH_2(CO_3)_4F_4$	deliquesce			10
Bariandite	$V_2O_4 \cdot 4V_2O_5 \cdot 12H_2O$	$-nH_2O$			11
Bassetite	$Fe^{+2}(UO_2)_2(PO_4)_2 \cdot 8H_2O$	$-nH_2O$			12
Bayleyite	$Mg_2(UO_2)(CO_3)_3 \cdot 18H_2O$	$-nH_2O$			13
Bianchite	$(Zn,Fe^{+2})SO_4 \cdot 6H_2O$	$+1H_2O \rightarrow 7$	59	20	c14
		$-1H_2O \rightarrow 6$	56	20	c14
Bieberite	$CoSO_4 \cdot 7H_2O$	deliquesce	94	20	2
		$-1H_2O \rightarrow 6$	70	20	2
Bischofite	$MgCl_2 \cdot 6H_2O$	deliquesce	33	20	c4
		$-2H_2O \rightarrow 4$	3	25	15
Bonattite	$CuSO_4 \cdot 3H_2O$	$+2H_2O \rightarrow 5$	33	20	c16
		$-2H_2O \rightarrow 1$	22	25	c16
Boothite	$CuSO_4 \cdot 7H_2O$	$-nH_2O$			17
Borax	$Na_2B_4O_5(OH)_4 \cdot 8H_2O$	deliquesce	99	20	2
		$-5H_2O \rightarrow 3$	50	20	c5
Boussingaultite	$(NH_4)_2Mg(SO_4)_2 \cdot 6H_2O$	deliquesce	96	25	2
		$-2H_2O \rightarrow 4$	20	25	c18
Boyleite	$(Zn,Mg)SO_4 \cdot 4H_2O$	$-nH_2O$			19
Brushite	$CaHPO_4 \cdot 2H_2O$	deliquesce	95	20	20
		$-nH_2O$			21
Buetschliite	$K_2Ca(CO_3)_2$	deliquese			22
Cadwaladerite	$Al(OH)_2Cl \cdot 4H_2O$	deliquesce			23
Carnallite	$KMgCl_3 \cdot 6H_2O$	deliquesce			17
		$-nH_2O$	3	32	c5
Carnotite	$K_2(UO_2)_2(VO_4)_2 \cdot 3H_2O$	$-nH_2O$			24
Carobbiite	KF	deliquesce	30	25	15
Chalcanthite	$CuSO_4 \cdot 5H_2O$	deliquesce	97	25	c16
		$-2H_2O \rightarrow 3$	33	25	c16
Chalcocyanite	$CuSO_4$	$+1H_2O \rightarrow 1$	3	25	c25
Chloraluminite	$AlCl_3 \cdot 6H_2O$	deliquesce	~40	25	15
Chlormanganokalite	K_4MnCl_6	deliquesce			21
Coquimbite	$Fe^{+3}_2(SO_4)_3 \cdot 9H_2O$	deliquesce	~73	20	2
		$-nH_2O$			26
Cyanochroite	$K_2Cu(SO_4)_2 \cdot 6H_2O$	$-2H_2O \rightarrow 4$	36	25	c27
Darapskite	$Na_3(SO_4)(NO_3) \cdot H_2O$	deliquesce			17
Delrioite	$CaSrV_2O_6(OH)_2 \cdot 3H_2O$	$-3H_2O \rightarrow$ meta			28
Dorfmanite	$Na_2HPO_4 \cdot 2H_2O$	$+5H_2O \rightarrow 7$	61	25	c29
		$-2H_2O \rightarrow 0$	28	25	c29
Douglasite	$K_2Fe^{+2}Cl_4 \cdot 2H_2O$	deliquesce			17
Epsomite	$MgSO_4 \cdot 7H_2O$	deliquesce	91	20	2
		$-1H_2O \rightarrow 6$	83	20	c5
Eriochalcite	$CuCl_2 \cdot 2H_2O$	deliquesce	68	20	30
Erythrosiderite	$K_2Fe^{+3}Cl_5 \cdot H_2O$	deliquesce			17
Ettringite	$Ca_6Al_2(SO_4)_3(OH)_{12} \cdot 26H_2O$	$-nH_2O$			21
Fairchildite	$K_2Ca(CO_3)_2$	deliquesce			21
Ferrarisite	$Ca_5H_2(AsO_4)_4 \cdot 9H_2O$	$-nH_2O$			31
Ferrinatrite	$Na_3Fe^{+3}(SO_4)_3 \cdot 3H_2O$	deliquesce			21
Ferrohexahydrite	$Fe^{+2}SO_4 \cdot 6H_2O$	$+1H_8O \rightarrow 7$	61	25	c14
Fibroferrite	$Fe^{+2}SO_4OH \cdot 5H_2O$				21
Franconite	$Na_2Nb_4O_{11} \cdot 9H_2O$	$-nH_2O$			32
Gaylussite	$Na_2Ca(CO_3)_2 \cdot 5H_2O$	$-nH_2O$			26

Mineral	Formula	Reaction			
Goslarite	$ZnSO_4 \cdot 7H_2O$	deliquesce	89	25	20
		$-nH_2O$			21
Gunningite	$(Zn,Mn)SO_4 \cdot H_2O$	$+5H_2O \rightarrow 6$	54	25	c25
		$-1H_2O \rightarrow 0$	4	25	c25
Halite	$NaCl$	deliquesce	75	20	33
Halotrichite	$Fe^{+2}Al_2(SO_4)_4 \cdot 22H_2O$	$-nH_2O$			26
Hanksite	$KNa_{22}(SO_4)_9(CO_3)_2Cl$	deliquesce	~75	RT	2
Heinrichite	$Ba(UO_2)_2(AsO_4)_2 \cdot 10\text{--}12H_2O$	$-nH_2O \rightarrow$ meta			34
Hellyerite	$NiCO_3 \cdot 6H_2O$	$-nH_2O$			35
Hexahydrite	$MgSO_4 \cdot 6H_2O$	$+1H_2O \rightarrow 7$	51	25	c14
		$-1H_2O \rightarrow 5$	41	25	c25
Hyrobasaluminite	$Al_4(SO_4)(OH)_{10} \cdot 12\text{--}36H_2O$	$-nH_2O \rightarrow 5$			36
Hydrodresserite	$BaAl_2(CO_3)_2(OH)_4 \cdot 3H_2O$	$-2H_2O \rightarrow 1$			37
Hydromagnesite	$Mg_5(CO_3)_4(OH)_2 \cdot 4H_2O$	$-nH_2O$			17
Hydrombobomkulite	$(Ni,Cu)_4Al_{16}(NO_3)_6(SO)_4$				
	$\cdot 13\text{--}14H_2O$	$-nH_2O \rightarrow 3$			38
Hydromolysite	$FeCl_3 \cdot 6H_2O$	deliquesce	~5	25	15
Hydroscarbroite	$Al_{14}(CO_3)_3(OH)_{36} \cdot nH_2O$	$-nH_2O \rightarrow 0$			39
Jokokuite	$MnSO_4 \cdot 5H_2O$	$+2H_2O \rightarrow 7$	84	15	c5
		$-4H_2O \rightarrow 1$	74	20	c5
Kafehydrocyanite	$K_4Fe^{+2}(CN)_6 \cdot 3H_2O$	$-3H_2O \rightarrow 0$	42	20	c5
Kainite	$MgSO_4 \cdot KCl \cdot 3H_2O$	deliquesce			17
		$-3H_2O \rightarrow 0$	3	32	c5
Kieserite	$MgSO_4 \cdot H_2O$	$+3H_2O \rightarrow 4$	21	25	c25
		$-1H_2O \rightarrow 0$	4	25	c25
Konyaite	$Na_2Mg(SO_4)_2 \cdot 5H_2O$	$-1H_2O \rightarrow 4$			40
Kremersite	$(NH_4,K)_2Fe^{+3}Cl_5 \cdot H_2O$	deliquesce			17
Langbeinite	$K_2Mg_2(SO_4)_3$	deliquesce			26
Lansfordite	$MgCO_3 \cdot 5H_2O$	$-2H_2O \rightarrow 3$		<~10	41
Laumontite	$CaAl_2Si_4O_{12} \cdot 4H_2O$	$-nH_2O$			26
Lawrencite	$(Fe^{+2},Ni)Cl_2$	deliquesce	56	25	15
Leonite	$K_2Mg(SO_4)_2 \cdot 4H_2O$	$+2H_2O \rightarrow 6$			21
Loeweite	$Na_{12}Mg_7(SO_4)_{13} \cdot 15H_2O$	deliquesce			21
Mallardite	$MnSO_4 \cdot 7H_2O$	$-2H_2O \rightarrow 5$		<9	42
Marthozite	$Cu(UO_2)_3(SeO_3)_3(OH)_2 \cdot 7H_2O$	$-nH_2O$			43
Mascagnite	$(NH_4)_2SO_4$	deliquesce	81	20	33
Matteuccite	$NaHSO_4 \cdot H_2O$	deliquesce	52	20	20
		$-1H_2O \rightarrow 0$	16	20	44
Melanterite[6]	$Fe^{+2}SO_4 \cdot 7H_2O$	deliquesce	95	20	2
		$-6H_2O \rightarrow 1$	57	20	2
Mendozite	$NaAl(SO_4)_2 11H_2O$	$-5H_2O \rightarrow 6$			21
Mercallite	$KHSO_4$	deliquesce	86	20	20
Meta-autunite	$Ca(UO_2)_2(PO_4)_2 \cdot 2\text{--}6H_2O$	$+nH_2O$	~38	20	i8
Metadelrioite	$CaSrV_2O_6(OH)_2$	$+3H_2O$			28
Metaschoderite	$Al_2(PO_4)(VO_4) \cdot 6H_2O$	$+2H_2O \rightarrow 8$			45
Metatyuyamunite	$Ca(UO_2)_2(VO_4)_2 \cdot 3\text{--}5H_2O$	$+nH_2O$	~55	24	c46
Meta-uranospinite	$Ca(UO_2)_2(AsO_4)_2 \cdot 8H_2O$	$+nH_2O$			47
Metavanuralite	$Al(UO_2)_2(VO_4)_2(OH) \cdot 8H_2O$	$+3H_2O \rightarrow 11$	47	20	48
Mirabilite	$Na_2SO_4 \cdot 10H_2O$	deliquesce	93	RT	20
		$-10H_2O \rightarrow 0$	81	25	c49
Molysite	$Fe^{+3}Cl_3$	deliquesce	~5	25	15
Monetite	$CaHPO_4$	$+2H_2O \rightarrow 2$			21
Monohydrocalcite	$CaCO_3 \cdot H_2O$	$-1H_2O \rightarrow 0$			50
Moorhouseite	$(Co,Ni,Mn)SO_4 \cdot 6H_2O$	$+1H_2O \rightarrow 7$	70	25	c51
Morenosite	$NiSO_4 \cdot 7H_2O$	deliquesce	93	25	2
		$-1H_2O \rightarrow 6$	87	25	c51
Nabaphite	$NaBa(PO)_4 \cdot 9H_2O$	$-nH_2O$			52
Nahpoite	Na_2HPO_4	$+2H_2O \rightarrow 2$	28	25	c29
Nastrophite	$Na(Sr,Ba)(PO_4) \cdot 9H_2O$	$-nH_2O$			53
Natrite	Na_2CO_3	$+1H_2O \rightarrow 1$	22	25	t54
Natrofairchildite	$Na_2Ca(CO_3)_2$	deliquesce			55
Natron	$Na_2CO_3 \cdot 10H_2O$	deliquesce	87	25	20
		$-3H_2O \rightarrow 7$	76	25	c56
Navajoite	$V_2O_5 \cdot 3H_2O$	$-nH_2O$			57

Table 3.2 *Continued*

Species	Formula[1]	Reaction[2]	%RH[3]	T[4]	Ref[5]
Nickelbischofite	$NiCl_2 \cdot 6H_2O$	deliquesce	~54	20	15
		$-2H_2O \rightarrow 4$	43	20	c5
Nitre	KNO_3	deliquesce	92	25	33
Nitratite	$NaNO_3$	deliquesce	74	20	c58
Nitrobarite	$Ba(NO_3)_2$	deliquesce	99	20	2
Nitrocalcite	$Ca(NO_3)_2 \cdot 4H_2O$	deliquesce	54	19	30
		$-1H_2O \rightarrow 3$	6	20	c5
Nitromagnesite	$Mg(NO_3)_2 \cdot 6H_2O$	deliquesce	52	25	c59
		$-4H_2O \rightarrow 2$	2	25	c59
Novacekite	$Mg(UO_2)_2(AsO_4)_2 \cdot 12H_2O$	$-nH_2O$			7
Olympite	Na_3PO_4	$+H_2O$			60
Paranatrolite	$Na_2Al_2Si_3O_{10} \cdot 3H_2O$	$-1H_2O \rightarrow 2$			61
Pascoite	$Ca_3V_{10}O_{28} \cdot 17H_2O$	$-nH_2O$			21
Pentahydrite	$MgSO_4 \cdot 5H_2O$	$+1H_2O \rightarrow 6$	41	25	c5
		$-1H_2O \rightarrow 4$	37	25	c5
Phaunouxite	$Ca_3(AsO_4)_2 \cdot 11H_2O$	$-1H_2O \rightarrow 10$			19
Phosphorroesslerite	$MgHPO_4 \cdot 7H_2O$	$-4H_2O \rightarrow 3$			21
Piypite	$K_2Cu_2O(SO_4)_2$	deliquesce			62
Pirssonite	$Na_2Ca(CO_3)_2 \cdot 2H_2O$	$-nH_2O$			17
Poitevinite	$(Cu,Fe^{+2},Zn)SO_4 \cdot H_2O$	$+2H_2O \rightarrow 3$	22	25	16
		$-1H_2O \rightarrow 0$	3	25	c25
Potassium alum	$KAl(SO_4)_2 \cdot 12H_2O$	$-6H_2O \rightarrow 6$	9	20	c63
Probertite	$NaCaB_5O_7(OH)_4 \cdot 3H_2O$	$-nH_2O$			26
Quenstedtite	$Fe_2^{+3}(SO_4)_3 \cdot 10H_2O$	$-1H_2O \rightarrow 9$			64
Rauvite	$Ca(UO_2)_2V^{+5}_{10}O_{28} \cdot 16H_2O$	$-nH_2O$			21
Retgersite	$NiSO_4 \cdot 6H_2O$	$+1H_2O \rightarrow 7$	84	20	c14
Rhomboclase	$HFe^{+3}(SO_4)_2 \cdot 4H_2O$	$-nH_2O$			26
Rinneite	$K_3NaFe^{+2}Cl_6$	deliquesce			26
Rokuhnite	$Fe^{+2}Cl_2 \cdot 2H_2O$	$+2H_2O \rightarrow 4$			65
Rossite	$CaV_2O_6 \cdot 4H_2O$	$-2H_2O \rightarrow 2$			21
Sal-ammoniac	NH_4Cl	deliquesce	79	20	c58
Sanderite	$MgSO_4 \cdot 2H_2O$	$+2H_2O \rightarrow 4$	22	31	c5
		$-1H_2O \rightarrow 1$	8	31	c5
Sasaite	$(Al,Fe^{+3})_{14}(PO_4)_{11}(SO_4)$ $(OH)_7 \cdot 83H_2O(?)$	$-nH_2O$			66
Scacchite	$MnCl_2$	deliquesce	~56	25	15
Schertelite	$(NH_4)_2MgH_2(PO_4)_2 \cdot 4H_2O$	deliquesce			67
Schoderite	$Al_2(PO_4)(VO_4) \cdot 8H_2O$	$-2H_2O \rightarrow 6$			45
Schoepite	$UO_3 \cdot 2H_2O$	$-nH_2O$			68
Sideronatrite	$Na_2Fe^{+3}(SO_4)_2(OH) \cdot 3H_2O$	$-nH_2O$			69
Sincosite	$CaV^{+4}_2(PO_4)_2(OH)_4 \cdot 3H_2O$	$-nH_2O$			21
Sinjarite	$CaCl_2 \cdot 2H_2O$	$+2H_2O \rightarrow 4$	11	25	c16
Sodium alum	$NaAl(SO_4)_2 \cdot 12H_2O$	$-6H_2O \rightarrow 6$	86	20	c63
Starkeyite	$MgSO_4 \cdot 4H_2O$	$+1H_2O \rightarrow 5$	37	25	c25
		$-3H_2O \rightarrow 1$	21	25	c25
Struvite	$(NH_4)MgPO_4 \cdot 6H_2O$	$-nH_2O$			26
Swartzite	$CaMg(UO_2)(CO_3)_3 \cdot 12H_2O$	$-nH_2O$			13
Sylvite	KCl	deliquesce	85	20	c4
Szmikite	$MnSO_4 \cdot H_2O$	$+4H_2O \rightarrow 5$	83	25	c70
		$-1H_2O \rightarrow 0$	17	20	c5
Szomolnokite[6]	$Fe^{+2}SO_4 \cdot H_2O$	$-1H_2O \rightarrow 0$	11	20	c5
		$+6H_2O \rightarrow 7$	57	20	2
Tachyhydrite	$CaMg_2Cl_6 \cdot 12H_2O$	deliquesce			17
Tamarugite	$NaAl(SO_4)_2 \cdot 6H_2O$	$+6H_2O \rightarrow 12$	86	20	c63
Tarapacaite	K_2CrO_4	deliquesce	88	20	20
Thenardite	Na_2SO_4	$+10H_2O \rightarrow 10$	81	25	c49
Thermonatrite	$Na_2CO_3 \cdot H_2O$	$+6H_2O \rightarrow 7$	66	25	t54
		$-1H_2O \rightarrow 0$	22	25	t54
Tincalconite	$Na_2B_4O_5(OH)_4 \cdot 3H_2O$	$+5H_2O \rightarrow 8$	50	20	c5
Tolbachite	$CuCl_2$	$+2H_2O \rightarrow 2$			71
Torbernite	$Cu(UO_2)_2(PO_4)_2 \cdot 8-12H_2O$	$-nH_2O \rightarrow meta$			72

Trona	$Na_3(CO_3)(HCO_3)\cdot 2H_2O$	$-nH_2O$			17
Tschermigite	$(NH_4)Al(SO_4)_2\cdot 12H_2O$	$-6H_2O \rightarrow 6$	7	25	c63
Tyuyamunite	$Ca(UO_2)_2(VO_4)_2\cdot 5-8H_2O$	$-nH_2O \rightarrow meta$	~55	24	c46
Uranocircite	$Ba(UO_2)_2(PO_4)_2\cdot 12H_2O$	$-nH_2O$			7
Uranopilite	$(UO_2)_6(SO_4)(OH)_{10}\cdot 12H_2O$	$-nH_2O$			21
Uranospathite	$HAl(UO_2)_4(PO_4)_4\cdot 40H_2O$	$-nH_2O$			7
Urea	$CO(NH_2)_2$	deliquesce	77	25	c58
Vandendriesscheite	$PbU^{+6}{}_7O_{22}\cdot 12H_2O$	$-nH_2O$			68
Vanuralite	$Al(UO_2)_2(VO_4)_2(OH)\cdot 11H_2O$	$-3H_2O \rightarrow meta$	47	20	48
Voltaite	$K_2Fe^{+2}{}_5Fe^{+3}{}_4(SO_4)_{12}\cdot 18H_2O$	$-nH_2O$			26
Zaherite	$Al_{12}(SO_4)_5(OH)_{26}\cdot 20H_2O$	$-nH_2O$			73
Zellerite	$Ca(UO_2)(CO_3)_2\cdot 5H_2O$	$-2H_2O \rightarrow meta$			74
Zinc-melanterite	$(Zn,Cu,Fe^{+2})SO_4\cdot 7H_2O$	deliquesce	89	25	20
		$-1H_2O \rightarrow 6$	65	25	c14

1 Formulas given are, when possible, taken from Fleischer[94].

2 The number of water molecules specified as being gained or lost in a reaction is for the reaction as written with the formula given. If the reaction is deliquescence, then, decomposition, oxidation, or hydrolysis may also occur.

3 The relative humidity at which a reaction occurs is dependent on temperature. The relative humidity given in column 4 applies at the temperature given in column 5. In cases where solid solution replacement of anions is common the stability limit given is that of the synthetic analogue of the end member containing only the anion shown in bold type in the formula.

4 Temperatures are given in °C. RT = Room Temperature (normally ~22 C); < = mineral is only stable below the temperature given. Below that temperature the mineral is subject to humidity-related phase transitions.

5 *References* (listed separately at end of this chapter)
  c = RH calculated from vapour pressure data in the reference given, using the relation per cent RH = $p_{H2O}/p^0_{H2O} \times 100$. Values of p^0_{H2O} used are from Weast[95] p.D–180.
  i = RH calculated from a vapour pressure interpolated from data in the reference given. Interpolation was done by linear regression performed on the relation $Log(p) = M \times (1/T) + B$.
  t = RH calculated from a vapour pressure that was calculated from thermodynamic data.

6 Values for the %RH for the efflorescence of melanterite and hydration of szomolnokite differ from those quoted by Waller (1987)[75] and referred to in this chapter. Experimental work subsequent to the writing of this chapter and of Waller (1987)[75] has shown that the values estimated from thermodynamic and solubility data are incorrect and should be replaced by the values listed here.

should, from thermodynamic considerations, result in hydration. This is probably the result of difficulty in nucleating the higher hydrate. For this reason hydration most commonly occurs in fine-grained material, and particularly in material that has formed by efflorescence of the higher hydrate. Repeated hydration/dehydration steps can result in decrepitation of specimens.

3.2.6 Specific stability limits

Table 3.2 lists mineral species which are known to undergo humidity-related phase transitions at RH levels between 5% and 90%. This list is by no means complete. Unfortunately the stability of mineral species at room temperature is seldom considered in any meaningful way when new minerals are being described. Consequently species are described as being stable when in fact they just happened to be stable in the observer's laboratory during the time that they were being studied. It is estimated that at least twice as many species as listed in Table 3.2 are in fact potentially unstable in museum environments.

The specific stability limits given are those of the equivalent synthetic compound. Except in cases where solid solution occurs, these limits are considered accurate to within 2% RH. If the limit given is marked as being approximate (i.e. ~), then the estimated accuracy is ±5% RH.

Numerous methods have been employed to determine the equilibrium vapour pressure of hydrate pairs and salt/solution pairs. Details of the most traditional methods, for example the tensimetric, isopiestic and gas current methods, can be found in many texts on physical chemistry. Other methods that are of interest in mineralogy because of their ability to work with small samples include vacuum microbalance[19] and x-ray diffraction[20] techniques in conjunction with temperature- and water-vapour-pressure control.

3.2.7 Prevention of humidity-related phase transitions

Consider Figure 3.10 again. It can be seen that at room temperature there are three distinct conditions at which mirabilite can be preserved: (1) at the upper phase boundary, that is with mirabilite in equilibrium with its saturated solution, (2) at the lower phase boundary, that is with mirabilite in equilibrium with thenardite, or (3) in the range between these points, that is by enforcing an RH at which mirabilite and water vapour are the only stable phases.

The first manner of preservation–storing a specimen in equilibrium with its saturated solution–is not advisable, since the solution phase permits ion mobility. This will lead to changes in crystal morphology. In certain cases oxidation and/or hydrolysis may also occur, as was described in Section 3.2.3. For these reasons, storage at mineral-saturated solution equilibrium conditions should only be considered for the preservation of minerals which are essentially insoluble. Examples of such minerals would be paranatrolite and laumontite.

The second manner of preservation is storing specimens in a sealed container at hydrate pair or hydrate-anhydrate equilibrium conditions. To avoid duplication, this section will consider only the case of preventing efflorescence. All arguments are similar if preventing hydration is the concern, except that the flow of water vapour to be avoided would be into rather than out of the container. In practice, using mirabilite as an example, the dry (meaning lacking any deliquesced water) specimen is sealed into a container. If this were done at an ambient 20°C and 50% RH, the mirabilite would effloresce, giving thenardite and water vapour until the RH within the container reached 75%. Unless the specimen is very small and is placed in a very large container, this establishment of equilibrium relative humidity (ERH) is not very taxing. In this example only 8 µg of mirabilite will effloresce for every 1 ml of air space. Once the hydrate pair equilibrium conditions have been established, assuming a rigid container and hence constant volume, it will persist indefinitely, perturbed only by temperature changes.

The long-term effect of diurnal temperature fluctuations on hermetically sealed containers will depend on whether the compensatory reaction of the specimen occurs within previously reacted areas or at the interface of reacted material and original crystal. In most cases previously reacted material is much more finely divided, and would probably react much more rapidly than non-reacted material. Reaction at the interface of reacted material and original crystal would only be expected in cases where the crystal structures of the two hydrates are highly conformal, and in fact where the transition occurs without decrepitation.

The third manner of preservation is to store the specimen at temperature and relative humidity conditions that are within the stability limits of all minerals in the specimen. This generally requires enclosing a substance or mixture that will enforce an appropriate RH level in a container. This method overcomes the basic problem of the previous two methods, which is that the mineral is in dynamic equilibrium with another phase and, hence, will chemically react to offset any gain or loss of water vapour to or from the container and any change in RH brought on by temperature change. While this method appears ideal in theory, in practice there are several technical difficulties which are usually surmountable but in some cases will result in the method of hermetic sealing of hydrate pairs being preferred. These technical difficulties will be discussed in Section 3.4.

3.3 Technical Considerations

3.3.1 Containers

Essential to any method of preventing interactions between specimens and atmospheric moisture (or the lack thereof) is some type of container. The container may be as large as the building in which a collection is housed, or as small as a coating of lacquer on a microscopic crystal. Regardless of size, the effectiveness of the container as a water vapour barrier must be known before the degree of protection it can offer can be assessed.

3.3.2 Lacquers

Lacquer coatings on specimens are essentially form-fitting containers. Their suitability as water-vapour barriers is limited by both the permeability of a continuous film of the lacquer and the presence of cracks or pinholes in the film. The limitations resulting from permeability are easy to calculate from available data.

One of the most impermeable thermoplastics suitable for use as a lacquer is polyvinylidene chloride (PVDC), with a permeability coefficient[21] (for water vapour) of $5 \times 10^{-11} cm^3$ (STP) cm cm^{-2} s^{-1} (cm Hg)$^{-1}$. Assuming a 1 cm^3 cube of mirabilite has a continuous 10 µm coating of PVDC and is stored at 25°C and 50% RH, it will dehydrate completely within 4.8×10^9 s, or about 150 years.

Most commonly used lacquers have permeability coefficients 100 to 10,000 times greater than this, and hence would allow such a hypothetical specimen to dehydrate completely in less than two years. In addition, the efflorescence of most minerals causes a smaller loss of water vapour than the efflorescence of mirabilite. This gives the mineral a shorter lifetime when exposed to equivalent conditions. For example, a 1 cm^3 cube of bieberite, $CoSO_4 \cdot 7H_2O$, which effloresces by losing 1 mole H_2O per mole solid (6.4 wt.%), coated with PVDC as described above, and stored at 30% RH below its ERH, would be completely effloresced within 23 years. In this case the commonly used lacquers would permit the specimen to completely effloresce in two months or less.

If permeability were the only consideration, it might be tempting to use heavier coatings to achieve better protection. For example, a 1 mm thick coating of PVDC would, in theory, permit a part of the 1 cm^3

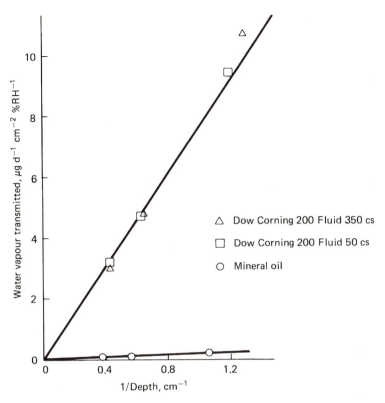

Figure 3.13 Water vapour transmission as a function of depth for silicone fluids and mineral oil at room temperature (25°C).

cube of mirabilite to exist unaltered for 15,000 years. Unfortunately it is unlikely that any lacquer would itself last anywhere near that long. In addition, lacquer coatings are generally not perfectly continuous, but contain microscopic to sub-microscopic cracks, voids and canals which permit gross leakage of water vapour. The extent to which these leaks will increase the transmission of water vapour through the coating is difficult to predict, as their formation depends on many factors. Consequently, the use of lacquers for the long-term preservation of specimens stored in an area where the average RH is not within the specimen's stability range is not recommended.

3.5.3 Jars and vials

Glass jars and vials with metal, plastic, glass or cork lids, covers or stoppers are commonly used containers. The efficiency of these containers varies widely with the type of closure employed and, in many cases, between individual examples of the same type of container.

Special precautions taken to ensure a good seal at the container-lid junction can markedly improve the performance of any of these types of containers. Nevertheless these are all imperfect containers, and, like lacquers, will permit the eventual total transition of a mineral if the average RH surrounding the container is not within the mineral's RH stability range.

3.3.4 Hydrophobic fluids

Hydrophobic fluids are liquids that are immiscible with water. They have at times been recommended for protecting specimens from inappropriate ambient RH levels. The water-vapour transmission rates of several hydrophobic liquids have been determined by storing glass vials containing various depths of each fluid over a layer of water, in a desiccating cabinet containing 'Drierite', and measuring weight losses. Because the data for the two 'Dow Corning 200' fluids were identical within experimental error, these data were combined (Figure 3.13). The water vapour transmission rate for these two fluids (at 25°C) is 7.7 µg cm⁻¹ d⁻¹ (%RH)⁻¹. This is rather high and indicates that these fluids would not be particularly effective for preserving minerals at average ambient RH levels outside the mineral's stability range. For example, a 1 cm³ crystal of mirabilite stored under 5 cm of either of these fluids in a 1 cm diameter vial at an average ambient RH of 50% would be completely effloresced after 74 years. As with lacquers, minerals having a dehydration step which involves a smaller weight percentage of water loss would alter completely within a shorter time.

The mineral oil, on the other hand, is a rather effective barrier, the measured water vapour transmission rate (at 25°C) being just 0.16 µg cm⁻¹d⁻¹ (%RH)⁻¹. A trace of the mirabilite described above would remain unaltered as long as 3,500 years when stored under mineral oil in a vial as described above. Nevertheless specimens will alter if hydrophobic fluids are depended on for protection. Figure 3.14 shows bieberite cleavage blocks after 17 years of storage in mineral oil, at an average ambient RH of ⁻30–35%.

Generally, if it is possible, it is better to store research material in hermetically sealed ampoules rather than in vials under oil.

Figure 3.14 Cleavage blocks of synthetic bieberite after 17 years of storage in oil. NMC #17287. Diameter of vial ~1 cm. The upper portion of the sample has effloresced, producing the paler six hydrate, moorhouseite.

3.3.5 Hermetic containers

Of all the materials suitable for the construction of containers, only glass and metal are impermeable. Even with these materials, however, obtaining a perfect seal at the closure is often difficult. One exception to this does exist – a glass ampoule or tube which can easily be hermetically sealed with a torch.

Both the closing and opening of these containers are simple and rapid. The quality of the seal can readily be ascertained by coating the ampoule with a leak-detection compound or detergent solution, then placing it in a bell jar or vacuum desiccator and evacuating. Any leaks present will result in an easily seen stream of bubbles. Ampoules are readily available from laboratory equipment suppliers in sizes from 1 to 50 ml. Larger sizes can be made from glass tubing, but care should be taken that the point where the tube is to be sealed is sufficiently removed from the sample, so that the sample is not heated during the sealing operation. It is also important that the ampoule be evacuated during the sealing process to prevent water driven off the heated part of the tube condensing on the specimen side of the closure point, since this could result in a change of conditions from hydrate transition ERH to mineral/solution ERH.

The construction of large hermetically sealed containers is possible,[22] but may prove too expensive for all but the most important mineral specimens. Padfield, Burke and Erhardt[23] have described the construction of a very well sealed container made from aluminium, glass and a viton O-ring.

3.4 Controlled climates

There are practical limitations to the size of containers that can be hermetically sealed. If a hermetically sealed container cannot be provided, then the third manner of preservation – storage within the mineral stability limits – can be employed. This requires the enforcement of a RH level within the specimen stability limits. The small size and highly individual needs of mineral specimens rule out mechanical or dynamic methods of RH control, and indicate the use of static methods.

The most popular methods for the static control of relative humidity in laboratories are through the use of saturated-salt solutions or sulphuric-acid solutions. Neither of these can be recommended for RH control in containers with mineral specimens, owing to the risk of contamination of the mineral specimens, through, for example, accidental agitation of the container. Silica gel, which is widely used in conservation for the control of RH, does not carry this risk of specimen contamination. It is dry to touch, even at high RH levels, and in the absence of strong agitation is essentially dust-free. Silica gel may be used to provide either a buffered or an enforced microclimate.

3.4.1 Buffered microclimates

Providing a buffered microclimate is appropriate for the preservation of mineral specimens when the average ambient RH level is within the stability range of the mineral specimen. A specimen is kept in a container at an average RH equal to the average ambient RH, but with the RH range encountered mitigated to the extent that it is within the stability range of the specimen. It is possible to calculate the efficiency of any particular container–buffer system,[24,25] but in cases where individual mineral specimens are to be protected, the amount of buffer for a given container volume can easily be made so large that sufficient buffering will certainly result. When enclosing specimens in a jar, it is not inconvenient to have a volume of silica gel equal to the volume of air. This gives a ratio of buffer to container volume well in excess of ten times the amount recommended by Thomson.[24] Still, it is essential that the average annual RH within the collection area be known and be within the stability range of the specimen. The average annual interior RH will, in most cases, differ considerably from the annual average exterior RH, and should be determined by hygrometric monitoring rather than estimated from exterior climatic data.

3.4.2 Enforced microclimates

By enforcing, within a specimen container, a relative humidity level that is within the stability limits of the specimen, preservation can be achieved. This is in a sense total preservation, in that the specimen will not have to react to offset any adverse ambient RH levels. Unfortunately the method suffers from numerous obvious and not so obvious practical problems.

The first and most major of these is that the stability limits of the specimen must be known. While this may seem an obvious problem, it is in some cases more of a subtle one, in that the alteration of even a minor constituent of a specimen can effect total destruction of the specimen. Consequently it is essential that all potentially humidity-sensitive species within a specimen be identified and their RH stability limits be known or determined. The importance of this consideration of even minor constituents cannot be overemphasized. To be safe, one should consider all minerals possible within a given paragenetic environment to be present in every specimen from that environment, unless it can be firmly established that any of these minerals are not present in the specimens. Even with a complete knowledge of the mineral species present in a specimen, knowledge of the stability limits of all these species may be lacking. The stability limits for a number of RH sensitive mineral species are given in Table 3.2. In cases where not all stability limits are known, new measurements will be required.

Once specimen stability limits have been determined, the technical considerations left are the choice of a container, a buffer, a method of conditioning the buffer, and a procedure for assembly and maintenance.

3.4.3 Containers

In the interest of low maintenance, containers should be chosen to be as impermeable as possible. In addition, since the specimen will be enclosed in a well-sealed container for an indefinite period of time, the container should not have any internal components which may break down, generating pollutants capable of reacting with the specimen. Commercially available jars are the obvious choice for containers for smaller specimens. For larger specimens carefully sealed aquariums might be considered. Again, however, it would be wise to consider the possible capacity of the sealant for pollutant emission. Generally transparent or at least partly transparent containers are preferable – an obvious necessity for display specimens, and psychologically more satisfying for reference specimens. It is reassuring to be able to see that the specimen is in fact surviving.

3.4.4 Buffers

The most essential consideration in choosing a buffer is that it must not be capable of contaminating or altering the specimen, either directly, or indirectly through the generation of reactive pollutants. Further, if employed in high RH situations, it should not support biological activities such as mould growth. Silica gel, which is chemically inert, dry to touch, and relatively inexpensive, is an acceptable RH buffer. Its capacity to buffer relative humidity derives from the facts that its moisture content varies greatly with the relative humidity to which it is in equilibrium, and that it responds rapidly by absorbing or desorbing moisture if the relative humidity in adjacent air rises or falls.

The buffering capacity of several types of silica gels have been determined.[26,27] The results for a regular and an intermediate density silica gel, both of which are useful for mineralogical preservation applications, are given in Figure 3.15. The greater the slope of the equilibrium moisture content (EMC) vs. RH curve, the greater the buffering capacity the gel will have. Figure 3.16 shows that in different RH ranges different silica gels will behave as the most effective buffer. In the interest of low maintenance requirements, the silica gel employed should have the highest possible EMC vs. RH slope in the RH range that it will be used to enforce. Consequently a regular density silica gel is used to enforce RH levels lower than 60% and an intermediate density silica gel is used to enforce RH levels higher than 60%.

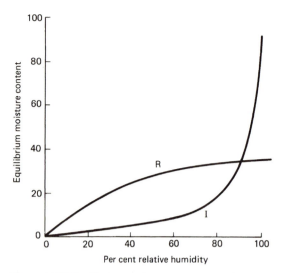

Figure 3.15 Equilibrium moisture content as a function of relative humidity for a regular density (R) and an intermediate density (I) silica gel. Based on data in Weintraub.[26]

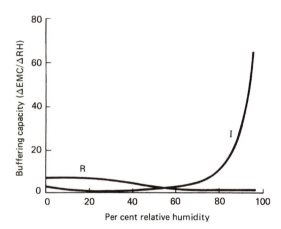

Figure 3.16 Buffering capacity of a regular density (R) and an intermediate density (I) silica gel as a function of relative humidity. Based on data in Weintraub.[26]

3.4.5 Conditioning the buffer

The most convenient method for conditioning small quantities of silica gel to a specific relative humidity, and hence equilibrium moisture content, is to allow them to equilibrate with a saturated salt solution which reliably enforces a known relative humidity. The ERH of a number of saturated salt solutions are given in Table 3.3. They are most dependable when prepared as a slurry, or sludge of excess salt in the saturated solution. In this state concentration gradients in the solution are minimized. The use of plastic

Table 3.3 Equilibrium relative humidities for saturated salt solutions suitable for conditioning silica gel. (All values from Greenspan – Table Reference 73)

Salt/temp. °C	20°C	25°C
Lithium bromide	6.6	6.4
Lithium chloride	11.3	11.3
Potassium acetate	23.1	22.1
Magnesium chloride	33.1	32.8
Potassium carbonate	43.2	43.2
Magnesium nitrate	54.4	52.9
Potassium iodide	69.9	68.9
Sodium chloride	75.5	75.3
Potassium chloride	85.1	84.4
Potassium sulphate	97.6	97.3

containers for solutions will result in less creeping of the solution than the use of glass or metal containers. The silica gel should be spread in a thin layer and enclosed in a relatively airtight container, together, but not in contact, with the saturated solution. The silica gel should be weighed every two or three days until a constant weight is reached.

Most silica gels are subject to hysteresis, i.e. the desorption curve is offset from the adsorption curve. A result of this is that a silica gel which has been following an adsorption curve and is then subjected to a lower RH will desorb water, following an isotherm which crosses the hysteresis loop. This is shown in Figure 3.17. This reduced EMC v RH slope results in a reduced capability to enforce

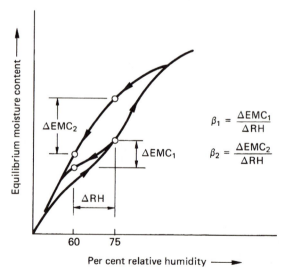

Figure 3.17 A hypothetical hysteresis loop illustrating the difference in buffering capacity (β) between 75 per cent and 60 per cent RH for a silica gel conditioned up to 75 per cent RH (β_1) and a silica gel conditioned down to 75 per cent RH (β_2).

RH within a particular range when the tendency is for the EMC to drift in a direction opposite to that in which the silica gel was conditioned. Weintraub[26] provides a more complete description of this sort of reduced RH-enforcing capacity. The practical result of this fact is that silica gel to be used to enforce an RH above the average ambient RH should be first conditioned to a too high RH, then conditioned downward to the required RH. Conversely a silica gel to be used to enforce an RH below the average ambient RH should be first conditioned to a too low RH, then conditioned upward to the required RH.

3.4.6 Assembly and maintenance

There are several general requirements concerning materials placed within the container. If the enforced RH is higher than 75% no organic material should be included. At such high humidities mould growth may occur. All hygroscopic materials that are going to be included in the container should be conditioned to the required RH level. If they are not, they may alter the EMC of the silica gel and, hence, the RH level enforced, in a manner that is usually difficult to predict quantitatively. The specimen itself will often fall into this hygroscopic substance category. This will be the case if a significant amount of the specimen has deliquesced, effloresced or hydrated, or if the matrix contains appreciable amounts of clay minerals, zeolites, or other minerals with variable water contents. Still another consideration, and one that cannot be overstressed, is that no material which desorbs or generates pollutants should be placed within the container. This last requirement precludes the use of wood, rubber, or similar specimen supports (Padfield *et al.*).[16]

Once the container, specimen, and silica gel have been selected and prepared, but before assembly, the silica gel should be tested to determine its buffering capacity. This is done by taking a carefully weighed 10g sample of the conditioned silica gel and exposing it to an RH just within the specimen stability limit towards which the container will drift. For example, the silica gel for a specimen which effloresces at 70% RH might be taken down to 75% RH for the test, while that for a specimen which hydrates at 30% RH might be brought up to 25% RH for the test. From the results of this test a buffering capacity in terms of weight percentage of the conditioned silica can be calculated. With this, and with a knowledge of the weight of silica gel included in the container and the final weight of the container with silica gel and specimen, a critical weight above or below which the container must not be allowed to drift can be calculated, using the formula:

$$CW = W_{total} + (W_{silica\ gel} \times W_{change}/W_{sample})$$

where CW = critical weight
W_{total} = weight of container + silica gel + specimen,
$W_{silica\ gel}$ = weight of conditioned silica gel in the completed specimen container,
W_{change} = weight lost or gained in buffering capacity test (– for loss, + for gain),
W_{sample} = weight of sample used for the buffering capacity test.

Maintenance will then call for periodic weighing of containers to ensure they have not reached the calculated critical weight. Once the critical weight is closely approached or has been reached, the silica gel should be replenished with a similar quantity of newly conditioned silica gel. Although containers may be designed to be sufficiently sealed and contain enough silica gel to last decades without reconditioning, it is recommended that routine weighings be performed on an annual basis. One of the reasons for a routine weighing is that it is always possible for a seal to be inadvertently damaged. In addition, without a routine it is too easy for the need of a specimen to be forgotten, especially if there is a change of personnel responsible for collection maintenance.

In summary, the only acceptable means for the long-term preservation of minerals subject to humidity-related phase transitions are hermetic sealing at a higher–lower hydrate ERH, or by enforcing an RH within the specimen stability limits. The fact that the former method compared to the latter required less than one tenth of the space to describe it is adequate testimony to the fact that it is far simpler to execute, and hence much less prone to human error. At the present time hermetic sealing is recommended over enforced microclimate storage whenever it can be applied. When enforced microclimates are employed, they should be designed only after careful consideration of all mineral species that might occur within the specimen.

3.5 Zeolitic water

Zeolitic water is loosely bound water of crystallization that can be lost from a mineral without the mineral undergoing a structural change. The zeolite mineral edingtonite is an example of a mineral that can lose a significant amount of zeolitic water at room temperature (Figure 3.18). This loss is reversible, and, hence, maintaining minerals having zeolitic water in a fully hydrated state is not necessary for their preservation. It is important to realize, however, that changes in the extent of hydration in these minerals does depend on the relative humidity of their storage environment, and that the results of

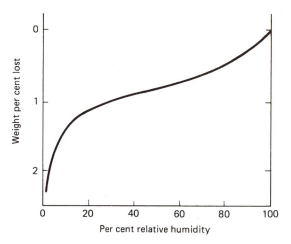

Figure 3.18 The variation of water content in the zeolite mineral edingtonite as a function of relative humidity.

water analyses and of physical property determinations performed on these minerals will vary accordingly.[28]

3.6 Hygrostatic specimens

Hygrostatic minerals have water contents that vary as functions of the ambient relative humidity, and swell as they absorb, and shrink as they desorb water. If a mineral specimen is able to alter its size and shape to accommodate this shrinkage and expansion, then a single drying cycle, if sufficiently gradual, is unlikely to cause great damage. If, however, the specimen cannot respond freely, then structural damage will occur as a result of a single drying cycle.

Radial clusters of vivianite crystals are examples of specimens that break up as a result of shrinkage due to dehydration of a clay-containing matrix between individual vivianite crystals (Figure 3.19). The compressive strength of the vivianite crystals effectively resists radial shrinkage. The inability of specimens to shrink radially results in the development of circumferential tensile stress. This, in turn, results in failure of the vivianite crystals across the perfect {010} cleavage (Figure 3.20).

Preservation methods for such anisotropic-hygrostatic specimens have not yet been developed. Storage at elevated levels of relative humidity would be the most obvious approach, but work must be done to determine what levels are required, and whether or not other forms of deterioration will occur as a result of maintaining damp, aerobic conditions. Other approaches might include freeze-drying or replacement of water with another substance that would maintain the original volume but would have less tendency to volatilize.

Figure 3.19 A radiating group of vivianite crystals from Anloua, Cameroons, NMNS #41027; width of specimen as viewed is 27 cm.

Figure 3.20 A detail of the specimen shown in Figure 3.19. Field of view is 6 cm. Note the separation of crystals along the {010} cleavage.

Certain mineralogical and petrological specimens that are capable of surviving a single drying cycle may still be damaged by exposure to repeated cycles of high and low RH. Among petrological specimens, the most likely to be affected are specimens that contain clays, particularly but not necessarily only expanding clays such as montmorillonite. As these specimens are exposed to high and low RH levels, they will expand and contract respectively, and

Figure 3.21 A system of mud cracks is a common example of fracture perpendicular to a surface as a result of tensile stress in a surface layer due to dehydration. (Photograph by P.F. Hoffmann. Reproduced with permission of Canadian Museum of Nature.)

Figure 3.22 A crazed surface of an opal specimen. CMN #20255; field of view is 16 × 12 mm. The development of crizzling in opals is directly analogous to the formation of mud cracks in a dry lake bottom.

these movements will cause weakening at grain boundaries, essentially producing stress fatigue.[29]

In addition, hygrostatic specimens exposed to a single rapid change in relative humidity may fail, owing to stresses associated with that single change.

For example, when a hygrostatic specimen is exposed to a decreased ambient relative humidity, the outermost part of the specimen will dry and consequently shrink more rapidly than the inner parts. The resultant tensile stress, if severe enough, will cause a network of fractures perpendicular to the specimen surface (Figure 3.21). Crazing in opal is an example of this behaviour (Figure 3.22). Conversely, when a hygrostatic specimen is exposed to increased ambient relative humidity, the outer layers will expand, causing compressional stress in outer layers, and fractures at 45° or less from the surface. These stresses and modes of failure are directly analogous to those experienced by a specimen undergoing thermal shock.

Unfortunately there are insufficient data available to make precise statements about what constitutes acceptable magnitudes and rates of change of relative humidity, but certainly exposure of susceptible specimens to large and rapid changes in relative humidity should be avoided.

References

Coverage of the literature is considered nearly complete to the date when the manuscript was submitted.

1 PARSONS, A.L., 'The preservation of mineral specimens', *American Mineralogist*, 7, pp. 59-63 (1922)
2 DEAN, J.A., *Lange's Handbook of Chemistry*, 12th edition McGraw-Hill Book Company, New York (1978)
3 BONNIER, C., 'Contribution a l'étude des carbonates ammoniacaux. Première partie, Décomposition par la chaleur du bicarbonate d'ammoniaque', *Annales de Chimie*, 5, pp. 46-52 (1926)
4 WEAST, R.C. ed., *Handbook of Chemistry and Physics*, 7th edition Chemical Rubber Publishing Company, Cleveland (1976)
5 KIRK, R.E., OTHMER, D.E., GRAYSON, M. and DECKROTH, D., *Kirk-Othmer concise encyclopedia of chemical technology* John Wiley & Sons, New York (1985)
6 GARRELS, R.M. and CHRIST C.L. *Solutions, Minerals, and Equilibria* Freeman, Cooper & Company, San Francisco (1965)
7 HEKINIAN, R., CHAIGNEAU, M. and CHEMINEE, J.L. 'Popping rocks and lava tubes from the Mid-Atlantic Rift Valley at 36° N', *Nature*, 245, pp. 371-3 (1973)
8 ROEDDER, E., *Fluid Inclusions* (Reviews in Mineralogy, 12) Mineralogical Society of America, Washington. p. 261 (1984)
9 DWIGHT, H.E., 'Account of the Kaatskill Mountains', *American Journal of Science*, 2, pp. 11-29 (1820)
10 NEWNHAM, R.E., *Structure-Property Relations* Springer-Verlag, New York (1975)
11 BARON, H.G., 'Thermal shock and thermal fatigue', in *Thermal Stress* eds P.P. Benham and H. Ford, Sir Isaac Pitman & Sons Ltd, London, pp. 182-206 (1964)
12 ADDLESON, L., *Materials for Building, Volume 1, Physical and Chemical Aspects of Matter and Strength of Materials*, Iliffe Books, London (1972)
13 BOKMAN, W., Personal communication (1985)

14 HOWIE, F.M.P., 'Museum climatology and the conservation of palaeontological material', *Special Papers in Palaeontology*, 22, pp. 103-25 (1979)

15 STEGER, H.F., 'Oxidation of sulphide minerals VII. Effect of temperature and relative humidity on the oxidation of pyrrhotite', *Chemical Geology*, 35, pp. 281-95 (1982)

16 PADFIELD, T., ERHARDT, D. and HOPWOOD, W., 'Trouble in store', *Science and Technology in the Service of Conservation*, Preprints of the contributions to the Washington Congress, 3-9 September 1982, International Institute for Conservation, London, pp. 24-27 (1982)

17 RICE, D.W., CAPPELL, R.J., PHIPPS, P.B.P. and PETERSON, P., 'Indoor atmospheric corrosion of copper, silver, nickel, cobalt, and iron', *Atmospheric Corrosion, Hollywood, FL, Oct 5-10, 1980, Proceedings*, ed. Ailor, W.H. John Wiley and Sons, Inc., Somerset, NJ, pp. 651-66 (1982)

18 PABST, A., 'Synthesis, properties, and structure of $K_2Ca(CO_3)_2$, buetschliite', *American Mineralogist*, 59, pp. 353-8 (1974)

19 HAMAD, S. EL D., 'A study of the reaction $Na_2SO_4 \cdot 10H_2O \rightarrow Na_2SO_4 + 10H_2O$ in the temperature range 0 to 25°C', *Thermochimica Acta*, 17, pp. 85-96 (1976)

20 EHLERS, E.G. and STILES, D.V., 'Melanterite–rozenite equilibrium', *American Mineralogist*, 50, pp. 1457-61 (1965)

21 YASUDA, H. and STANNETT, V. 'Permeability coefficients', in *Polymer Handbook*, eds J. Brandrup and E.H. Immergut, 2nd edition John Wiley & Sons, New York, III-229-40 (1974)

22 PASSAGLIA, E., BROWN, D. and DICKENS, B. *The Preservation of the Constitution of Puerto Rico*, National Bureau of Standards, NBSIR 83-2743, (1983)

23 PADFIELD, T., BURKE, M. and ERHARDT, D., 'A cooled display case for George Washington's Commission', *Preprints*, ICOM Committee for Conservation, 7th Triennial Meeting, Copenhagen, 84.17.38-42 (1984)

24 THOMSON, G., 'Stabilization of RH in exhibition cases: hygrometric half-time', *Studies in Conservation*, 22, pp. 85-102 (1977)

25 BRIMBLECOMBE, P. and RAMER, B., 'Museum display cases and the exchange of water vapour', *Studies in Conservation*, 28, pp. 179-88 (1983)

26 WEINTRAUB, S., 'Studies on the behavior of RH within an exhibition case. Part I: Measuring the effectiveness of sorbents for use in an enclosed showcase', *Preprints*, ICOM Committee for Conservation 6th Triennial Meeting, Ottawa, 81/18/4, pp. 1-11 (1981)

27 LAFONTAINE, R.H., *Silica Gel*, Technical Bulletin No. 10 Canadian Conservation Institute, National Museums of Canada, Ottawa (1984)

28 GRICE, J.D., GAULT, R.A. and ANSELL, H.G. 'Edingtonite: the first two Canadian occurrences', *Canadian Mineralogist*, 22, pp. 253-8 (1984)

29 HUDEC, P.P., 'Standard engineering tests for aggregate: What do they actually measure?', in *Decay and Preservation of Stone*, ed. Winkler, E.M., Engineering Geology Case Histories Number 11 Geological Society of America, Boulder, CO, pp. 3-6 (1978)

References for Table 3.2

1 HAWLEY, G.G., *The Condensed Chemical Dictionary*, 8th edition, (Van Nostrand Reinhold Company, New York, (1971)

2 Author's data.

3 FANG, J.H. and ROBINSON, P.D. 'Alunogen, $Al_2(H_2O)_{12}(SO_4)_3 \cdot 5H_2O$: Its atomic arrangement and water content', *American Mineralogist*, 61, pp. 311-17 (1976)

4 ACHESON, D.T. 'Vapor pressures of saturated aqueous salt solutions', *Humidity and Moisture. 3. Fundamentals and Standards*, eds A. Wexler and W.A. Wildhack, Reinhold Publishing Company, New York, pp. 521-30 (1965)

5 RANDALL, M. 'Free energy of chemical substances, activity coefficients, partial molal quantities, and related constants', *International Critical Tables*, ed. E.W. Washburn, Volume VII, McGraw-Hill Book Company, Inc., New York, pp. 224-353 (1930)

6 WILLIAMS, S.A. 'Anthonyite and calumetite, two new minerals from the Michigan copper district', *American Mineralogist*, 48, pp. 614-19 (1963)

7 WALENTA, K. 'Uranospathite and arsenuranospathite', *Mineralogical Magazine*, 42, pp. 117-28 (1978)

8 BEINTEMA, 'On the composition and the crystallography of autunite and the meta-autunites', *Recueil des Travaux Schimiques du Pays-Bas et de la Belgique*, 57, pp. 155-75 (1938)

9 KELLER, P., personal communication (1980)

10 KHOMYAKOV, A.P., KUROVA, T.A., NECHELYUSTOV, G.N. and PILOYAN, G.O. 'Barentsite $Na_7AlH_2(CO_3)_4F_4$ – a new mineral' (in Russian), *Zapiski Vsesoyuznogo Mineralogicheskogo Obshchestva*, 112 (1983), pp. 474-9 (not seen; extracted from *Mineralogical Abstracts*, 35, #84M/1925) (1984)

11 CESBRON, F. and VACHEY, H. 'La bariandite, nouvel oxyde hydrate de vanadium (IV) et (V)', *Bulletin de la Société Française de Minéralogie et de Cristallographie*, 94, pp. 49-54 (1971) (not seen; extracted from *Mineralogical Abstracts*, 22, #71-2329) (1971)

12 FRONDEL, C. 'Bassetite and uranospathite', *Mineralogical Magazine*, 30, pp. 243-53 (1954)

13 AXELROD, J.M., GRIMALDI, F.S., MILTON, C. and MURATA, K.J. 'The uranium minerals from the Hillside Mine, Yavapai County, Arizona', *American Mineralogist*, 36, pp. 1-22 (1951)

14 BONNELL, D.F.R. and BURRIDGE, L.W. 'The dissociation pressures of some salt hydrates', *Faraday Society, Transactions*, 31, pp. 473-8 (1935)

15 RICHARDSON, G.M. and MALTHUS, R.S. 'Salts for the static control of humidity at relatively low levels', *Journal of Applied Chemistry*, 5, pp. 557-67 (1955)

16 COLLINS, E.M. and MENZIES, A.W.C. 'A comparative method for measuring aqueous vapor and dissociation pressures, with some of its applications', *Journal of Physical Chemistry*, 40, pp. 379-97 (1936)

17 PARSONS, A.L. 'The preservation of mineral specimens', *American Mineralogist*, 7, pp. 59-63 (1922)

18 CAVEN, R.M. and FERGUSON, J. 'The dissociation pressures of hydrated double sulphates. Part II. Various double

sulphates of the type M⁺⁺SO₄,M⁺₂SO₄·6H₂O', *Journal of Chemical Society, London*, 125, pp. 1307-12 (1924)

19 BARI, H., CATTI, M., FERRARIS, G., IVALDI, G. and PERMINGEAT, F., 'Phaunouxite, Ca₃(AsO₄)₂.11H₂O, a new mineral strictly associated with rauenthalite', *Bulletin de la Société Française de Minéralogie et de Cristallographie*, 105, pp. 327-32 (1982) (not seen; extracted from *American Mineralogist*, 68, p. 850) (1983)

20 O'BRIEN, F.E.M. 'The control of humidity by saturated salt solutions', *Journal of Scientific Instruments*, 25, pp. 73-6 (1948)

21 PALACHE, C., BERMAN, H. and FRONDEL, C. *The System of Mineralogy*, Volume II, 7th ed., John Wiley and Sons, Inc., New York (1951)

22 PABST, A. 'Synthesis, properties and structure of K₂Ca(CO₃)₂, buetschliite', *American Mineralogist*, 59, pp. 353-8 (1974)

23 GORDON, S.G. 'Results of the Chilean mineralogical expedition of 1938. Part V. - Cadwaladerite, a new aluminum mineral from Cerro Pintados, Chile', *Notulae Naturae of The Academy of Natural Sciences of Philadelphia*, 80 (1941)

24 HILLEBRAND, W.F. 'Carnotite and tyuyamunite and their ores in Colorado and Utah', *American Journal of Science*, Fifth series, 7, pp. 201-16 (1924)

25 FOOTE, H.W. and SCHOLES, S.R. 'The vapour pressure of hydrates, determined from their equilibria with aqueous alcohol', *Journal of the American Chemical Society*, 33, pp. 1309-26 (1911)

26 BANNISTER, F.A. 'The preservation of minerals and meteorites', *Museums Journal*, 36, pp. 465-76 (1937)

27 CAVEN, R.M. and FERGUSON, J. 'The dissociation pressures of hydrated double sulphates. Part I. Hydrated cupric alkali sulphates', *Journal of the Chemical Society*, London, 121, pp. 1406-14 (1922)

28 SMITH, M.L. 'Delrioite and metadelrioite from Montrose County, Colorado', *American Mineralogist*, 55, pp. 185-200 (1970)

29 WATERFIELD, C.G. and STAVELEY, L.A.K. 'Thermodynamic investigation of disorder in the hydrates of disodium hydrogen phosphate', *Transactions of the Faraday Society*, 63, pp. 2349-56 (1967)

30 HEDLIN, C.P. and TROFIMENKOFF, F.N. 'Relative humidities over saturated salts in the temperature range from 0-90°F', *Humidity and Moisture. 3. Fundamentals and Standards*, eds A. Wexler and W.A. Wildhack, Reinhold Publishing Company, New York, pp. 519-20 (1965)

31 BARI, H., PERMINGEAT, R., PIERROT, R. and WALENTA, K. 'La ferrarisite Ca₅H₂(AsO₄)₄.9H₂O, une nouvelle espèce minérale dimorphe de la guerinite', *Bulletin de la Société Française de Minéralogie et de Cristallographie*, 103, pp. 533-40 (1980)

32 JAMBOR, J.L., SABINA, A.P., ROBERTS, A.C., BONARDI, M., RAMIK, R.A. and STURMAN, B.D. 'Franconite, a new hydrated Na-Nb oxide mineral from Montreal Island, Quebec', *Canadian Mineralogist*, 22, pp. 239-43 (1984)

33 WEXLER, A. and HASEGAWA, S. 'Relative humidity–temperature relationships of some saturated salt solutions in the temperature range 0° to 50°C', *Journal of Research of the National Bureau of Standards*, 53, pp. 19-26 (1954)

34 GROSS, E.B., COREY, A.S., MITCHELL, R.S. and WALENTA, K. 'Heinrichite and metaheinrichite, hydrated barium uranyl arsenate minerals', *American Mineralogist*, 43, pp. 1134-43 (1958)

35 WILLIAMS, K.L., THREADGOLD, I.M. and HOUNSLOW, A.W. 'Hellyerite, a new nickel carbonate from Heazlewood, Tasmania', *American Mineralogist*, 44, pp. 533-8 (1959)

36 SUNDERMAN, J.A. and BECK, C.W. 'Hydrobasaluminite from Shoals, Indiana', *American Mineralogist*, 54, pp. 1363-73 (1969)

37 JAMBOR, J.L., SABINA, A.P. and STURMAN, B.D. 'Hydrodresserite, a new Ba-A1 carbonate from a silicocarbonatite sill, Montreal Island, Quebec', *Canadian Mineralogist*, 15, pp. 399-404 (1977)

38 MARTINI, J.E.J. 'Mbobomkulite, hydrombobomkulite, and nickelalumite, new minerals from Mbobo Mkulu cave, eastern Transvaal', *Annals, Geological Survey South Africa*, 14, pp. 1-110

39 DUFFIN, W.J. and GOODYEAR, J. 'A thermal and x-ray investigation of scarbroite', *Mineralogical Magazine*, 32, pp. 353-62 (1960)

40 VAN DOESBURG, J.D.J., VERGOUWEN, L. and VAN DER PLAS, L. 'Konyaite, Na₂Mg(SO₄)₂.5H₂O, a new mineral from the Great Konya Basin, Turkey', *American Mineralogist*, 67, pp. 1035-8 (1982)

41 LANGMUIR, D. 'Stability of carbonates in the system MgO-CO₂-H₂O', *Journal of Geology*, 73, pp. 730-54 (1965)

42 LINKE, W.F. *Solubilities of inorganic and metal organic compounds*, Vol. II, American Chemical Society, Washington (1965)

43 CESBRON, F., OOSTERBOSCH, R. and PIERROT, R. 'Une nouvelle espèce minérale: la marthozite. Uranyl-seleite de cuivre hydrate', *Bulletin de la Société Française de Minéralogie et de Cristallographie*, 92, pp. 278-83 (1969) (not seen; extracted from *Mineralogical Abstracts*, 21, #70-751) (1970)

44 HEJTMANKOVA, J. and CERNY, C. 'Solid-gas equilibrium in the binary system sodium hydrogen sulphate-water', *Collection of Czechoslovak Chemical Communications*, 39, pp. 1787-93 (1974)

45 HAUSEN, D.M. 'Schoderite, a new phosphovanadate mineral from Nevada', *American Mineralogist*, 47, pp. 637-48 (1962)

46 STERN, T.W., STIEFF, L.R., GIRHARD, M.N. and MEYROWITZ, R. 'The occurrence and properties of metatyuyamunite, Ca(UO₂)₂ (VO₄)₂·3-5H₂O', *American Mineralogist*, 41, pp. 187-201 (1956)

47 WALENTA, K. 'Die sekundaren Uranmineralien des Schwarzwaldes', *Jahresheft geol. Landesamt Baden-Wurttemberg*, 3, pp. 17-51 (1958) (not seen; extracted from *American Mineralogist*, 45, p. 254) (1960)

48 CESBRON, F. 'Nouvelles données sur la vanuralite. Existence de la meta-vanuralite', *Bulletin de la Société Française de Minéralogie et de Cristallographie*, 93, pp. 242-8 (1970) (not seen; extracted from *Mineralogical Abstracts*, 21, #70-3425) (1970)

49 WILSON, R.E. 'Some new methods for the determination of the vapor pressure of salt-hydrates', *Journal of the American Chemical Society*, 43, pp. 740-25 (1921)

50 SKINNER, H.C.W., OSBALDISTON, G.W. and WILNER, A.N. 'Monohydrocalcite in a guinea-pig bladder stone, a novel occurrence', *American Mineralogist*, 62, pp. 273-7 (1977)

51 SCHUMB, W.C. 'The dissociation pressures of certain salt

hydrates by the gas-current saturation method', *Journal of the American Chemical Society*, 45, pp. 342-54 (1923)

52 KHOMYAKOV, A.P., KOROBITSYN, M.F., MEN'SHIKOV, Y.P. and POLEZHAEVA, L.I. 'Nabaphite, NaBaPO$_4$.9H$_2$O, a new mineral', *Doklady Akademii Nauk SSSR*, 266, pp. 707-10 (1982) (not seen; extracted from *Mineralogical Abstracts*, 34, #83M/3664) (1983)

53 KHOMYAKOV, A.P., KAZAKOVA, M.E., POPOVA, G.N. and MALINOVSKII, Y.A. 'Nastrophite, Na(Sr,Ba)PO$_4$.9H$_2$O, a new mineral' (in Russian), *Zapiski Vsesoyuznogo Mineralogicheskogo Obshchestva*, 110, pp. 604-7 (1981) (not seen; extracted from *American Mineralogist*, 67, p. 857) (1982)

54 WAGMAN, D.D., EVANS, W.H., PARKER, V.B., SCHUMM, R.H., HALOW, I., BAILEY, S.M., CHURNEY, K.L. and NUTALL, R.L. 'The NBS tables of chemical thermodynamic properties: selected values for inorganic and C$_1$ and C$_2$ organic substances in SI units', *Journal of Physical and Chemical Reference Data*, 11, Supplement #2 (1982)

55 KAPUSTIN, Y.L. *Mineralogy of Carbonatites*, Nauka Publishers, Moscow (1971); English translation by Amerind Publishing Co. Pvt. Ltd., New Delhi, p. 157 (1980)

56 BAXTER, G.P. and COOPER, W.C. 'The aqueous pressure of hydrated crystals. II. Oxalic acid, sodium sulfate, sodium acetate, disodium phosphate, barium chloride', *Journal of the American Chemical Society*, 46, pp. 923-33 (1924)

57 WEEKS, A.D., THOMPSON, M.E. and SHERWOOD, A.M. 'Navajoite, a new vanadium oxide from Arizona', *American Mineralogist*, 40, pp. 207-12 (1955)

58 EDGAR, G. and SWAN, W.O. 'The factors determining the hygoscopic properties of soluble substances. I. The vapor pressures of saturated solutions', *Journal of the American Chemical Society*, 44, pp. 570-7 (1922)

59 EWING, W.W., KLINGER, E. and BRANDNER, J.D. 'Studies on the vapor pressure-temperature relations and on the heats of hydration, solution and dilution of the binary system magnesium nitrate-water,' *Journal of the American Chemical Society*, 56, pp. 1053-57 (1934)

60 KHOMYAKOV, A.P., BYKOVA, A.V. and MALINOVSKII, Y.A. 'Olympite, Na$_3$PO$_4$, a new mineral' (in Russian), *Zapiski Vsesoyuznogo Mineralogicheskogo Obshchestva*, 109, pp. 476-9 (1980) (not seen; extracted from *Mineralogical Abstracts*, 32, #81-1874) (1981)

61 CHAO, G.Y. 'Paranatrolite, a new zeolite from Mont St-Hilaire, Quebec', *Canadian Mineralogist*, 18, pp. 85-8 (1980)

62 BERGASOVA, L.P., FILATOV, S.K., SERAFIMOVA, E.K. and STAROVA, G.L. 'Piypite K$_2$Cu$_2$O(SO$_4$)$_2$ - a new mineral of volcanic sublimates' (in Russian), *Doklady Akademii Nauk SSSR*, 266, pp. 707-10 (1982) (not seen; extracted from *American Mineralogist*, 70, pp. 437-8) (1985)

63 HEPBURN, J.R.I. and PHILLIPS, R.F. 'The alums. Part I. A study of the alums by measurement of their aqueous dissociation pressures', *Journal of the Chemical Society*, London, pp. 2569-78 (1952)

64 BANDY, M.C. 'Mineralogy of three sulphate deposits of northern Chile', *American Mineralogist*, 23, pp. 669-760 (1938)

65 HODENBERG, R. VON, and STRUENSEE, G. VON 'Rokuhnite, FeC1$_2$.2H$_2$O, a new mineral', *Neues Jahrbuch fur Mineralogie, Monatshefte*, pp. 125-30 (1980) (not seen; extracted from *American Mineralogist*, 66, p. 219 (1981)

66 MARTINI, J. 'Sasaite, a new phosphate mineral from West Driefontein Cave, Transvaal, South Africa', *Mineralogical Magazine*, 42, pp. 401-4 (1978)

67 FRAZIER, A.W., LEHR, J.R. and SMITH, J.P. 'The magnesium phosphates hannayite, schertelite and bobierrite', *American Mineralogist*, 48, pp. 635-41 (1963)

68 CHRIST, C.L. and CLARK, J.R. 'Crystal chemical studies of some uranyl oxide hydrates', *American Mineralogist*, 45, pp. 1026-61 (1960)

69 ZODROW, E.L., WILTSHIRE, J. and MCCANDLISH, K. 'Hydrated sulfates in the Sidney Coalfield of Cape Breton, Nova Scotia. II. Pyrite and its alteration products', *Canadian Mineralogist*, 17, pp. 63-70 (1979)

70 CARPENTER, C.D. and JETTE, E.R. 'The vapor pressures of certain hydrated metal sulfates', *Journal of the American Chemical Society*, 45, pp. 578-90 (1923)

71 BERGASOVA, L.P. and FILATOV, S.K. 'The new mineral tolbachite', *Doklady Akademii Nauk SSSR*, 270 pp. 415-417 (1983) (not seen; extracted from *Mineralogical Abstracts*, 35, #84M/1941) (1984).

72 VOCHTEN, R. and DOORSELAER, M.V. 'Secondary uranium minerals of the Cunha Baixa Mine', *Mineralogical Record*, 15, pp. 293-7 (1984)

73 RUOTSALA, A.P. and BABCOCK, L.L. 'Zaherite, a new hydrated aluminum sulfate', *American Mineralogist*, 62, pp. 1125-8 (1977)

74 COLEMAN, R.G., ROSS, D.R. and MEYROWITZ, R. 'Zellerite and metazellerite, new uranyl carbonates', *American Mineralogist*, 51, pp. 1567-78 (1966)

75 WALLER, R.R. *An experimental ammonia gas treatment method for oxidized pyritic mineral specimens*, International Council of Museums, Committee for Conservation, preprints of the Eighth Triennial Meeting, Sydney, pp. 625-30 (1987)

4

Elements, alloys and miscellaneous minerals

Frank M. Howie

Fewer than thirty elements are known to occur in native or uncombined form in the earth's crust, its atmosphere and in meteorites.[1] Of these, eight are gases, fourteen are metals and six are solid non-metals. Several of these are extremely uncommon but, of the solid elements, at least ten may deteriorate. In addition, several naturally occurring alloys are known (see Table 4.1). For the purposes of this chapter gases other than those found as inclusions in minerals are ignored.

4.1 Corrosion and oxidation of native elements and alloys

As would be expected, several of the so-called noble metals, including gold, platinum, osmium and iridium and their alloys and amalgams, are comparatively unaffected by normal chemical processes, and as a consequence tend to accumulate in certain rock types. Perhaps unexpectedly, reactive elements and alloys such as iron, iron-nickel alloys, sulphur, arsenic and bismuth have survived in unreacted forms, mainly as a result of formation and existence under anoxic conditions. These minerals are in fact difficult to maintain in good condition in the atmosphere.

The damaging constituents of the atmosphere are essentially water vapour (too much), carbon dioxide, sulphur and nitrogen oxides, organic acids, and, last but not least, dust. Oxygen of course plays an essential role in oxidation and corrosion reactions, but by itself is not damaging at normal storage or exhibition temperatures. Some corrosion or oxidation products may in fact be protective, for example the dull oxidized layer which rapidly develops on the surface of fresh copper, zinc and lead.

Dust in the atmosphere is ubiquitous and can cause severe problems in collections–not least by becoming indurated on surfaces over short periods of time, but also by encouraging corrosion and other reactions on surfaces on which it has settled. The average urban/industrial atmosphere contains anything between two and $1,000 \ mg/m^3$ of dust, which usually consists of carbon (soot etc.), organic compounds, sulphates, chlorides, metal oxides, silicates, spores, etc.[2]

Dust particles are damaging because they absorb moisture and acids, such as sulphuric, nitric, acetic and formic acids, from the air, forming potential electrolytes on the surface on which they settle. Dust particles on a surface can also form galvanic cells where the surface is either metallic or electrically conducting (e.g. many sulphides).[2] The higher the relative humidity, the greater is the potential for corrosion. Cool, dry, dust-free atmospheres are really essential for ensuring that susceptible elements and alloys remain in good condition.

Corrosion is usually perceived as a surface phenomenon peculiar to metals and alloys. It can occur in many ways and may be localized or uniform in its effects. Corrosion may be one of a number of other reactions occurring simultaneously on a surface. With native elements and alloys, atmospheric corrosion may be the most serious conservation problem. The rate of corrosion usually increases with increasing relative humidity, and for many metals and alloys critical humidity levels or bands above which corrosion occurs rapidly have been observed. The critical relative humidity bands for iron,[4] nickel[5] and copper[6] are approximately 65% (see Figure 4.1), 85% and 95–8% respectively. These levels are no more than indicators. Some materials corrode faster in sheltered indoor conditions than outside, where rain washes away corrosive reactants.

Pitting corrosion will occur on surfaces where dust particles settle. The mechanism responsible is a form of galvanic (electrochemical) corrosion, where

Table 4.1 Naturally occurring native elements

Non-metals	Problems
Antimony	Tarnish – grey
Argon	
Arsenic	Tarnish – dark grey
Helium	
Krypton	
Neon	
Nitrogen	
Oxygen	
Radon	(See p.115)
Selenium	
Sulphur	(See p. 30)
Tellurium	
Xenon	

Metals	Problems
Bismuth	Tarnish – pink
Copper	Tarnish, corrosion
Gold	
Indium	
Iron	Corrosion, oxidation
Iridium	
Lead	Tarnish
Mercury	Highly volatile
Nickel	Tarnish
Osmium	
Platinum	
Silver	Tarnish, sulphidization
Tin	Metastable at $13.2°C$
Zinc	

Some alloys	Problems
Algodonite (Cu_6As)	
Allemontite (AsSb)	Tarnish – grey/brown
Amalgam (HgAg)	
Austenite (Fe, C)	
Cementite (Fe_3C)	
Domeykite (Cu_3As)	Tarnish – grey/yellow/black
Dyscrasite (Ag_3Sb)	Tarnish – grey/yellow/black
Kamacite (Fe, Ni)	
Martensite (Fe_2C)	
Osmiridium (OsIr)	
Platiniridium (PtIr)	
Taenite (Fe, Ni)	

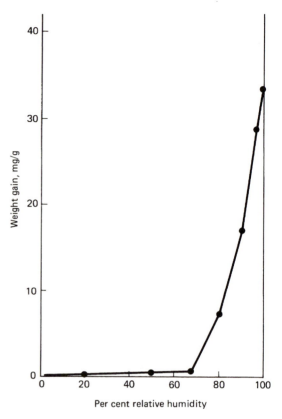

Figure 4.1 Weight change for iron exposed to air containing sulphur dioxide v relative humidity (adapted from Evans and Taylor 1972).

the part underneath the dirt particle is oxygen-deficient and corrodes anodically and the surrounding parts act cathodically, shedding electrons.[2] Corrosion under these circumstances is depicted in Figure 4.2(a–b).

Corrosion occurs most commonly with iron (steel), copper (see Figure 4.2) and nickel and some of their alloys. With single element specimens, two or more phases may occur together, for example the austenite/martensite/carbon/ferrite/cementite phases in smelted iron or steel. The different phases have different standard electrode potentials. In the presence of surface moisture with dust/electrolyte, galvanic pairs form. Those with the highest values

become anodes and those with the lowest become cathodes. At the anode iron goes into solution, and iron hydroxide, which further oxidizes in air to hydrated ferric oxides, is formed. At the cathode hydrogen ions are released and oxidized to water by atmospheric oxygen (see Figure 4.2.b). This is the basis for the rusting[4,7] of iron and steel, and this type of corrosion is also seen on meteorites (see Chapter 7). A related type of corrosion results from handling clean unoxidized metallic minerals, alloys, sulphides, etc. Skin deposits an electrolyte, i.e. sodium chloride in sweat, on the surface of the mineral, which results in the development of fingerprints, probably by galvanic corrosion.

The extent of corrosion always depends on the corrosivity of the environment, with acidic industrial and salt-laden marine atmospheres being far more damaging than simply humid atmosphere.[2] The effects of these types of atmosphere on mineral specimens have not been recorded to any great extent, but some collections in urban areas or those close to a seacoast show more marked damage than rural collections,

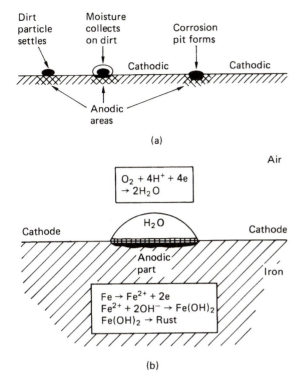

Figure 4.2 Corrosion can occur when:
(a) Dirt settles on a metallic surface and the oxygen concentration under the particle decreases, causing this part to corrode anodically.
(b) Moisture accumulates as droplets on the surface of iron, and the part beneath the droplet corrodes, anodically forming rust.

with corrosion indicators such as copper, silver and certain sulphides showing marked deterioration.

The use of conventional anti-corrosion treatments such as coatings or paints, stove-enamelling, hot dipping, electro-plating, and the use of corrosion-inhibitors, are options which are not available for preserving minerals. The only realistic option is environmental control.

4.2 Control of storage environment

Since it is not practical to prevent contact between oxygen and minerals susceptible to corrosion or oxidation and, as most of these processes are dependent on the presence of water vapour, together with dust, etc., the simplest method of control is to diminish levels of exposure to water vapour and other contaminants. At relative humidities of less than 30% corrosion of susceptible metals and alloys is negligible, except that surface dust may cause localized corrosion at 20% RH.

Humidity control can be achieved by dynamic or static methods.

4.2.1 Dynamic methods of humidity control

Air-conditioning for the removal of industrial contaminants and close control of relative humidity on the scale required for storerooms, exhibition galleries or buildings housing collections is costly and labour-intensive to maintain. With efficient computer-based environmental management systems, however, becoming readily available, and air-conditioning plant becoming easier to install, more and more institutions are opting for this form of environmental control. The subject is, however, outside the scope of the present work, and readers are referred to texts such as the *ASHRAE Handbook*.[8]

4.2.2 Static methods of humidity control

Static methods of control in collections and exhibitions all utilize enclosure of susceptible material with some type of moisture or contaminant adsorbent. Silica gel can be dried or conditioned to give a certain RH in an enclosure. Other adsorbents used include zeolites, molecular sieves and activated carbon. Adsorbents such as silica gel also remove other pollutants, such as ammonia, to some extent.

Enclosures can consist of small specimen jars, sealed drawers, complete cabinets and exhibit cases. These can be constructed of plastic, glass, metal or timber, or any combination of these materials. Certain plastics, metals and glass are preferable (see Appendix I).

Timber and related products, such as fibreboard, particleboard, etc., are potentially hazardous to specimens because their moisture-buffering characteristics above approximately 35% RH are poor. This, combined with the tendency for wood products to liberate organic contaminants such as acetic or formic acid, encourages corrosion and other types of deterioration.

4.3 Effects of acid vapours emitted by timbers and other materials on calcareous specimens

The corrosive effect of acetic and other organic acid vapours on metals has been known for centuries.[10] The corrosion of ferrous and non-ferrous metals and certain types of museum object by acetic and formic acids which are emitted by timbers is now well known.[11,12] Timbers such as oak, birch, and chestnut, and several of the timber-based compositions, can emit significant amounts of volatile organic acids, especially at high storage temperatures (20°C) and

Table 4.2 The effect of corrosive atmospheres on metals and alloys

Material	Effect of vapour: Formic acid (6ppm)	Acetic acid (1% solution in water	Distilled water (100% humidity)
Iron/Steel (0.1% carbon)	105 g/m²#	328 g/m²*	2 g/m²*
Copper	19 g/m²#	22 g/m²*	0 G/M²*
Nickel	2 g/m²#	9 g/m²*	0 g/m²*
Cadmium (99.5% pure)	14 g/m²#	99 g/m²*	60 G/M²*
Tin	0 g/m²#	0 g/m²*	0 g/m²*
Brass (20:30)	0 g/m²#	12 g/m²*	0 g/m²*
Aluminium (99% pure)	0 g/m²#	23 g/m²*	3 g/m²*
Silver	0 g/m²#	0 g/m²*	0 g/m²*

#Weight loss in gms per square metre of metal surface exposed to 6 ppm of formic acid vapour at 100 per cent RH for 21 days.
*Weight loss in gms per square metre of metal surface exposed to vapour above 1 per cent acetic acid in water at 25°C for 6 days.

relative humidities (60–70%). Kiln-dried timber is more likely than other types to be the source of acidic vapours, because the drying process destroys the wood hemicellulose, releasing acetyl groups which are easily hydrolysed to acetic acid.

Table 4.2 details some of the data and observations on the effects of corrosive atmospheres on metals and alloys. The data in this table can be applied direct to the storage of susceptible native elements and alloys in collections, although the climatic conditions may not be so aggressive. Long periods of continuous or intermittent exposure to even fractional ppm concentrations of acetic or formic acid can be damaging.

Timber is not the only source of corrosive vapours; adhesives, fabrics, paints, plastics, rubbers, etc. are all potential contributing factors[14]. Formaldehyde and other aldehydes may also contribute, either direct or because they readily oxidize in air to acids.

Fitzhugh and Gettens[12] and others have described the products of corrosive atmosphere on various types of objects in collections. Table 4.3 outlines the results of exposure to acid vapours from timber, mainly acetic and formic, on the mineral content of some museum objects. Detailed research into the efflorescences forming on shells[13] suggests that both the products and reaction mechanisms in this type of corrosion are clearly more complex than was thought.

The treatment of corroded specimens, whether metals or calcareous, has not been satisfactorily resolved. Thorough washing of affected material in distilled water to remove any potentially hygroscopic contaminants, or dry cleaning where material is water susceptible, is satisfactory. Subsequent specimen storage or exhibition should be in environments free of corrosive contaminants such as organic acids, hydrogen sulphide or sulphur dioxide, and where temperatures are kept below 20°C at relative humidities of 30% or less. For further information on corrosive contaminants, their sources and effects, see Appendix I.

Table 4.3 Effect of exposure of minerals to timber

Material	Corrosion products
Shells (calcite)	Hydrated calcium acetate/formate Calclacite
Limestone (pure)	Calclacite
Limestone (contaminated with KNO₃)	Hydrated calcium/chloride nitrate/acetate (see Figure 4.3)
Clay (baked) (calcium silicate)	Calclacite
Terracotta (calcium silicate)	Calclacite
Lead	Lead formate
Glass (high soda)	Sodium formate
Enamels	Sodium formate
Birds' eggs (calcium carbonate)	Calcium acetate

References

1 HEY, M.H. and EMBREY, P.G., *Chemical Index of Minerals* British Museum, Natural History, London (1975)

2 DIAMENT, R.M.F., *The Prevention of Corrosion* Business Books Ltd, London (1971)

3 LA QUE, F.L. and COPSON, H.P. (Eds), *Corrosion Resistance of Metals and Alloys*, ACS Monograph No. 158, Reinhold New York (1963)

4 EVANS, V.R. and TAYLOR, C.A.J., 'The mechanism of atmospheric rusting', *Corrosion Science*, 12, pp. 227–46 (1972)

5 VERNON, W.H.J., 'The fogging of nickel', *J. Inst. Metals*, 48, p. 121 (1932)

6 KAMMLOT, I., 'Atmospheric corrosion of copper', *Journal of the Electrochemical Society*, 24, pp. 26–36 (1984)

7 *The Influence of Salts in Rusts on the Corrosion of Underlying Steels* BISRA, London (1968)

8 American Society of Heating, Refrigeration and Air-Conditioning Engineers Inc., *ASHRAE Handbook, 1987, HVAC systems and applications* ASHRAE, Atlanta (1987)

9 THOMSON, G., *The Museum Environment* Butterworths, London (1985)

10 DONOVAN, P.D. and STRANGER, J., 'Corrosion of metals and their protection in atmospheres containing organic acid vapours', *Br. Corrosion J.*, 6, pp. 132-8 (1971)

11 VAN TASSEL, R., 'Une efflorescence d'acetatochlorure de calcium sur des roches calcaires dans des collections', *Bull. du Mus. Royal d'Histoire Nat. de Belgique*, 21, pp. 1-11 (1945)

12 FITZHUGH, E.W. and GETTENS, R.J., 'Calclacite and other efflorescent salts on objects stored in wooden museum cases', *Science and Archaeology*, Cambridge, Mass., pp. 91-102 (1971)

13 TENNANT, N.H. and BAIRD, T., 'The deterioration of Mollusca collections: identification of shell efflorescence', *Studies in Conservation*, 30, pp. 73-85 (1985)

14 BRIMBLECOMBE, P., 'Composition of Museum Atmospheres, *Atmospheric Environment*, 24B, pp. 1-8 (1990)

5

Sulphides and allied minerals in collections

Frank M. Howie

More than 500 distinct species of sulphides and allied minerals are known to occur naturally. Allied minerals here include selenides, tellurides, arsenides, antimonides, bismuthides, oxysulphides and the sulphosalts, i.e. sulpharsenites, sulphantimonites, sulpharsenates and sulphantimonates, sulphostannates, sulphogermanates, sulphobismuthites and sulphovanadates. These minerals form a more or less complete bridge between (1) the native elements and alloys and (2) complex oxygen-rich minerals such as sulphates. A few examples will illustrate this diversity.

Allemontite (AsSb) is strictly classified as a native alloy, although it behaves as a metal, an arsenide and an antimonide; pyrite (FeS_2) and cooperite (PtS) are simple sulphides; chabourneite ($Tl_8Pb_4Sb_{21}As_{19}S_{68}$) is one of the more complex sulphosalts; and coyoteite ($NaFe_3S_5(OH)_2$) is an esoteric oxy-sulphide. With such a diverse range of minerals it would obviously be no easy task to elucidate trends in stability.

The classification of sulphides suggested by Kostov et al,[1] based on metallic composition, is probably more amenable to stability and oxidation predictions than Dana's system,[2] which is based on crystallographic structure. A complicating factor with minerals in this grouping is that they frequently occur as mixtures or solid solutions of two or more species. In many cases stability is very much dependent on the intimacy of the species or the presence of small quantities of contaminant elements.

Until a few years ago[3] little had appeared in the literature on the solid state and surface chemistry of naturally occurring sulphides and allied minerals. Vaughan and Craig,[4] who described the chemistry of sulphides but not sulphosalt minerals, gave extensive fundamental data on thermochemistry and equilibria but little on the low temperature and surface oxidation reactions and transformations exhibited by sulphides and allied minerals.

Historical observation on the stability of minerals, including several sulphides, are found in ancient works by Aristotle, Theophrastus, Dioscorides, Pliny, Albertus Magnus and Agricola. Many of the early natural philosophers must have observed common sulphide transformations, such as pyrite to copper as (melanterite), realgar to orpiment or cinnabar to mercury, and these reactions may well have helped to originate some of the alchemical transmutation principles.[5] During the period 1650–1900 some of the common sulphide mineral oxidation reactions were described in considerable detail, for example marcasite and pyrite (see Chapter 6), and by the late 1800s some of the simpler mechanisms had been defined.

Work in the mid-1900s highlighted the extent of the activity of biological agents in the formation and oxidation of sulphides. More recently analytical techniques such as electron spectroscopy and sweep voltammetry have opened up new fields in the exploration of the surface reactivity of sulphides. Recent research has revealed some of the surface and near surface reaction mechanisms which precede or initiate oxidation of certain of the sulphides and allied minerals.[6,7,8] Beigler and Horne,[9] Vaughan et al.[10] and others have described phenomena such as photo-oxidation, tarnishing and the initial formation of oxygen-rich complexes which probably precede the destructive formation of oxygen-rich anions such as sulphates or arsenates.

There are, however, still considerable gaps in the understanding of the low-temperature chemistry of the sulphides and allied minerals when exposed to the atmosphere, whether under the relatively 'dry' and constant conditions in museum collections or under the fluctuating wet and dry conditions in mines, ore dumps and tailings.

The information presented in this chapter is a distillation of many sources for the professional and amateur mineralogist. This chapter is not intended to be comprehensive and if anything perhaps highlights the paucity of data and information in the field. Perhaps it will encourage others to fill in the gaps.

5.1 Reactivity and oxidation mechanisms of sulphides and allied minerals

The search for economic deposits of minerals and metals has, in turn, been accelerated and hampered by the unstable and metastable sulphides and allies which occur in igneous, sedimentary and metamorphic rocks.[11,12] The instability of some of these minerals and their oxidation in zones of weathering lead to the formation of acid percolating solutions which help to leach out valuable metals, either naturally, through the formation of secondary enriched deposits, or by human design. Many sulphides and sulphosalts are key sources of rare and essential metals and, indirectly, of sulphur and arsenic and their compounds.

The very mechanisms which are beneficial for extraction by leaching may, however, have the reverse effect where unwanted sulphide-rich deposits or gangues are uncovered during mining operations. Here sub-aerial sulphide oxidation, in some cases accelerated by a variety of micro-organisms, produce acidic pollution which adds considerably to the costs of operations by contributing to the corrosion of mining equipment. Environmentally, the desolate dumps, tailings and landfills from these mining operations often take decades to lose their acidity and become clothed with vegetation.[13]

Although seemingly remote from mineral processing activities, mineral collections are affected by many of the same reactions but on a smaller scale. Few collections of worth have remained unscathed by the havoc caused by pyrite oxidation (see Chapter 6). There are, however, a very large number of other sulphides and allied minerals which undergo reactions ranging from slight surface tarnish to complete disintegration through complex oxidation mechanisms. Unfortunately there are few useful ground rules for predicting the stability of these minerals in air. Chemistry, paragenesis and diagenesis, as well as environmental parameters, all appear to play roles in determining stability.

5.1.1 Oxidation mechanisms

Sulphide and allied minerals undergo a variety of transformations and oxidation reactions under normal conditions of storage or exhibition environment. A number of silver, copper, arsenic, antimony

and bismuth sulphides will tarnish extremely rapidly on exposure to light and air. Several references to these changes have been made in the literature: Palache *et al.*,[14] Pearl,[15] Parsons,[16] Bannister,[17] Pearl,[18] Sinkankas,[19] and Howie[20] give some data on many of the sulphides and allied minerals which are sensitive to light and/or oxygen. Little reference has been made to the part water vapour plays in many oxidation reactions, other than recommendations that 'high' relative humidity should be avoided.

In this section we shall consider three main types of low temperature instability:

- Photo-induced transformation and photo-oxidation.
- Tarnish development.
- Oxidation reactions which occur in the presence of water vapour in air.

5.1.2 Photo-induced transformations and oxidation reactions

There are a number of sulphides, sulphosalts and allied minerals which are affected to varying degrees by exposure to electromagnetic radiation, including light.[21] Acanthite and several other sulphide minerals, realgar and stibnite come into this group. Here the activation energy of the transformation

$$MS + hv \rightarrow M^0 + S^0 \text{ (or } S^{2-}) \qquad (1)$$

is very low, and sufficient radiation (hv is the energy quantum of visible, UV of shorter, wavelength radiation) is absorbed to form mobile electronic species which migrate through the crystal lattice until trapped at defect sites or lattice imperfections.[22] At the trap an 'F-centre' complex is formed when the defect is at or near a surface anion vacancy:

$$MS_x + e^- \rightarrow (M_{Sx})^{e^-} \qquad (2)$$

The electron associated with the complex is also associated with both the anion vacancy and production of a positive 'hole'. Where two or more electronic species home in on one trap, the following reaction probably occurs:

$$(MS_x)e^- + {}^-e(MS_x) \rightarrow 2x(S^{2-}) + 2M^0 \qquad (3)$$

The two electrons may be carried away as sulphur dioxide or (possibly) hydrogen sulphide, leaving the surface anion impefection. Further reactions at or near the site will lead to a build-up of metallic product. Many silver minerals undergo this type of decomposition (see later in this chapter and Chapter 3).

Some of these reactions may be reversible. Others are not, for example where the product from the reaction with polluting sulphur species such as hydrogen sulphide is not the same sulphide as that which constituted the original mineral.

5.1.3 Tarnish development in air

Where the sequences (1), (2) and (3) described above lead to the production of ionic species which react with oxygen, metastable or stable sulphide-oxygen and/or sulphide-oxygen-hydroxyl complexes will be produced on the surface. Just what part exposure to light plays in this type of reaction is not always clear although it is not always an essential factor:

$$MS_x \xrightarrow[O_2]{hv} (MS_{x-1}0_n) + S^0 \text{ (or } S^{2-} \text{ or } SO_2) \quad (4)$$

$$MS_{x-1}O_n \xrightarrow[O_2 + H_2O]{} MS_{x-1}O.OH + S^0 \text{ (or } S^{2-} \text{ or } SO_2) \quad (5)$$

$$MS_x \xrightarrow[O_2 + H_2O]{} MS_{x-1}.O.OH \quad (6)$$

$$MS_{x-1}O + O_2 + H_2O \rightarrow MS_{x-1}O.O.OH \quad (7)$$

Surface reactions resulting in tarnishing and the formation of superficial oxidation products may be extremely fast; Vaughan *et al.*[10] measured a period of minutes for the formation of the familiar 'peacock' iridescence on the surface of bornite. The development of dull tarnish on minerals such as pyrite or arsenopyrite, on the other hand, may take days, weeks or even months. Clearly there are different mechanisms at work, and most of these have not been elucidated as yet. Removal, or reversal of the tarnishing reaction, can sometimes be achieved, for example, by reduction with 'active' hydrogen.

Many sulphides and allied minerals can be classified as semiconductors.[23] Although little direct research has been carried out into possible links between surface instability effects and semiconductivity in minerals, considerable evidence has amassed in other fields, such as solid state electronics and alloy metallurgy, which suggests that investigations into mineral semiconductivity would be very rewarding.

Mitchell and Woods[24] investigated the 'oxidized' layer on tarnished pyrite, and concluded that neither physical adsorption nor chemisorption had occurred – rather that the layer was a true oxide, perhaps complexed with water of hydration or hydroxyl ions. This type of reaction probably occurs with a number of sulphide and sulphosalt minerals, including most of those containing iron, lead, iron-copper mixtures, copper, nickel and cobalt. One feature of this type of oxidation is that it imparts a high rest potential to the sulphide when immersed in acid solutions, i.e. 'passivation' occurs.

Reactions of the type described in Sections 5.1.2 and 5.1.3 are surface effects. Characteristically they are not influenced to any degree by crystal or particle size, shape or form.[25]

The oxidation reactions described in the next section are predominantly influenced by the extent of surface area available for oxidation. These may, in part at least, be initiated by reactions of the type described in this section. These oxidation reactions are normally destructive, with the non-metallic phase being irreversibly replaced by a new oxygenated anion.

5.1.4 Oxidation reactions which occur in the presence of water vapour

The difficulty of categorizing these reactions lies in defining the limits of water-vapour activity; at the lowest levels of water activity, i.e. relative humidity 0–30%, the reactions are probably self-limiting, proceeding as shown by equations (6) or (7) in Section 5.1.3. At the highest levels of water vapour activity (85% to 100% RH) it may be difficult to differentiate between atmospheric and aqueous oxidation. In the area between these limits lies the normal range of environmental conditions found in most mineral collection, storage and exhibition areas.

One of the major characteristics of this type of reaction is the oxidation of sulphide sulphur to a sulphate species and the retention of H^+ ions in the reactive aqueous multi-layer or film on the surface of the mineral.[26] Similar reactions occur with many arsenides and antimonides and with some bismuthides and sulphosalts.

The general reactions for low temperature sulphide (etc.) oxidation have been (non-stoichiometrically) represented classically thus:

$$MS + O_2 \rightarrow M^+ + SO_4^{2-} \quad (8)$$

$$MS + H_2O + O_2 \rightarrow MSO_4 + H_2SO_4 \quad (9)$$

Recently, however, it has been found that for many sulphides, these reactions are not simple oxidations but involve at least two separate mechanisms:

(a) Surface/air adsorption reactions, where unstable or metastable sulphide–oxygen–water species are produced.
(b) Electrochemical reactions producing sulphate and involving the reduction of atmospheric oxygen to water.

Reactions (8) and (9) involve, in the case of sulphides, the change from S^{2-} to S^{6+}, a net loss of eight electrons; some of these electrons will react with oxygen and water in the adsorbed layer on the sulphides surface thus:

$$O_2 + H_2O + 4e^- \rightarrow 4OH^-, \qquad (10)$$

the cathodic reduction of water. The balancing half-reaction is the anodic dissolution of the sulphide to sulphate with the production of protons:

$$MS + 4H_2O \rightarrow M^{2+} + SO_4^{2-} + H^+ + 8e^- \qquad (11)$$

There is increasing evidence that sulphur dioxide polythionates and thiosulphates (and their analogues of tellurium, selenium, etc.) are some of the initial products of oxidation of pyrite, galena, pyrrhotite and other sulphides, tellurides and selenides. This strongly supports the hypothesis that the metal–sulphur (or metal–tellurium or metal–selenium) bond breaks before the S–S (or Te–Te or Se–Se) bond, with production of transient polysulphide (or Se or Te analogue), which oxidizes to polythionate:[26]

$$MS \rightarrow M^+ \; S \qquad (12)$$

$$S \xrightarrow{O_2} Sx(transient) \rightarrow (S_xO_y)_n \qquad (13)$$

The polythionate anion will usually rapidly rearrange or disproportionate to sulphate.

Further oxidation of the cation, i.e. $M^{2+} \rightarrow M^{3+}$, will release further electrons for cathodic reduction and, in the cases of metal moieties such as ferric and cupric ion, these will effectively react with unchanged sulphide.

It is this phase of the oxidation which can bring in various bacteria, especially species of thiobacteria and sulphobacteria.[27] The part that these organisms play in metallic ore leaching is well-known. Their role, however, in the oxidation of sulphide minerals in collections is doubtful, because, unless the storage area is frequently flooded or perpetually saturated with high levels of water vapour, bacteria-inoculated sulphides will rapidly become too toxic and acidic to support bacterial oxidation.

5.1.5 Properties of some susceptible sulphides and care in collections

The sulphides described below are among those whose susceptibility to oxidation is fairly well documented. Some indications for conditions of storage are given, but neither the list of susceptible minerals nor the environmental data are by any means complete.

Where susceptible sulphides and other minerals and rocks are encountered in collections, documentation of individual specimen condition is fundamentally important. This should consist of, at the

Figure 5.1 (a) Acanthite showing grey friable crust formed during exposure to light. (b) Acanthite stored in dark, showing original lustre (courtesy of the Natural History Museum, London). ($\times 0.5$)

very least, a detailed description of the storage conditions, state of the specimens as found, extent of any alteration, condition photographs, date of examination and details of any treatment, proposed or tried. The removal of alteration products by physical or chemical means is only desirable if those products threaten the integrity of the specimen.

Acanthite

Acanthite is the low temperature form of silver sulphide (Ag_2S) and occurs naturally as a pseudomorph after the high-temperature form argentite. Acanthite rapidly tarnishes to lead-grey when exposed to artificial light,[28] sunlight,[29] UV light and cathode rays.[30] In very short periods of time the tarnish develops into a grey crust, which becomes loose and friable (Figure 5.1).

The solid-state reactions which acanthite may undergo include:

$$Ag_2S \xrightarrow[\text{room temp}]{\text{x-rays (in vacuo)}} Ag + S$$

$$Ag_2S \xrightarrow[\text{in air}]{\text{sunlight} \atop \text{or x-rays}} Ag + SO_2 \; (\text{or } H_2S)$$

$$Ag_2S \xrightarrow[\text{room temp}]{\text{water vapour}} \text{no reaction}$$

$$Ag_2S \xrightarrow[150°C +]{\text{water vapour}} Ag + H_2S$$

Stephens,[28] Guild[29] and Whitehead[31] describe in detail the photo-reactivity of acanthite to high-intensity tungsten light. The development of a fine network of dots on polished surfaces after 5–15 seconds' exposure to arc light was followed by etching or pitting. This reaction has not been observed to occur as rapidly during exposure to sunlight or diffuse exhibition lighting, however, over prolonged periods of time, surface degradation almost always occurs. The rapid photo-chemical reaction of polished silver-containing sulphide ore minerals during microscopic examination under plane-polarized light has caused misidentification of phases.

The conservation strategy for acanthite and other minerals which react similarly to light includes:

- Exclusion from light of all types.
- Keeping away from radioactive minerals, sources of x-rays.
- Keeping in dust-proof containers.
- Not exposing to contaminated atmospheres, i.e. avoiding proximity to oxidizing pyrite and other sulphides.
- Avoiding exposure to hydrogen sulphide and to the acidic by-products of case or cabinet construction (see Appendix I).

Exhibiting acanthite for anything but short periods will inevitably cause surface decomposition. Avoid sunlight, tungsten light, spotlights. Maximum light levels should be 50–100 lux, with UV-screened fluorescent lamps.

Stromeyerite and mckinstyrite

Both stromeyerite ($AgCuS$) and mckinstryite ($AgCuS$) appear to behave like mixtures of acanthite and chalcocite. Stromeyerite will alter in colour from dark grey to blue on exposed surfaces,[16] probably due to oxidation to metallic silver and covelline. For storage and exhibition apply the same strategy as for acanthite.

Chalcocite

Chalcocite (Cu_2S) is probably stable to light (see, however, Mellor[30]) and does not react appreciably with oxygen at low temperatures. Some specimens slowly develop a dull tarnish, and moist air turns amorphous chalcocite steel blue. Recent studies[33] indicate that self-limiting superficial oxidation will occur in samples contaminated with iron (as pyrite), possibly by oxidation to covelline. Chalcocite is sensitive to both acidic and alkaline atmospheres, and care should be taken to avoid exposure to materials which generate pollutants (see Appendix I).

Covelline

Covelline (CuS) is commonly contaminated with iron, and develops a blue to purple tarnish (see bornite and chalcopyrite for discussions on the destabilizing effect of iron on copper sulphides). Synthetic cupric sulphide is a reactive material which oxidizes in moist air exothermically.[30] The mineral covelline does not appear to share this characteristic.

Chalcopyrite

Chalcopyrite ($CuFeS_2$) usually exhibits a marked tarnish,[34] which ranges from brassy yellow to iridescent blue/red to dull brown (Plate 5.1). It has also been reported as oxidizing slowly to a mixture of copper and iron sulphates on exposure to moist air at low temperatures.

Recent work by Buckley and Woods,[33] Beigler and Horne[9] and others shows that chalcopyrite undergoes a series of complex oxidation and transformation reactions in which iron is transported to the surface of the mineral, where it is oxidized (possibly electrochemically) to a hydrated ferric oxide. The remaining sulphur-enriched copper sulphide (possibly CuS_2) is oxidized slowly to copper sulphate. The rate of oxidation is increased with decreasing grain size.

Ragusa and Madwick[35] showed experimentally that the oxidation of crushed chalcopyrite could be brought about at high relative humidity (70% to 80%) by *thiobacillus ferro-oxidans*. Such reactions are, however, unlikely to occur in mineral collections under the range of normal storage conditions.

Thorpe *et al.*[36] maintain that tarnish development in chalcopyrite may be intimately controlled by the extent of silver contamination. Chen and Petruk[37] suggested that the differing colours exhibited by tarnished chalcopyrite depended on its silver content. Chen *et al.*[34] suggested that the tarnish film consists of silver which reacts with sulphur from the surface of chalcopyrite to form acanthite.

Chalcopyrite should ideally be stored and exhibited at relative humidities of less than 50% and not be exposed to atmospheres polluted with acid or alkaline contaminants (see Appendix I). Chalcopyrite may be sensitive to light, although this has not been definitely confirmed.

Stannite (Cu_2FeSnS_4) similarly slowly tarnishes from grey-black to blue, and may undergo similar reactions to chalcopyrite. No stability data have been recorded.

Bornite

Palache[14] noted that bornite (Cu_5FeS_4) rapidly tarnished upon exposure to moist air to a purple iridescence (see Plate 5.2) and Buckley and Woods[38] described some of the possible reactions. Beigler and Horne,[9] Buckley and Woods[33] and Vaughan *et al.*[10] have investigated the surface oxidation reactions and development of tarnish on bornite. As with chalcopyrite, it appears that upon initial exposure to air the near-surface structure of bornite is destabilized and reconstructed. Iron diffuses to the surface, where it reacts with oxygen and moisture to produce hydrated iron oxide. This reaction appears to occur in a matter of minutes of exposure to air (see Figure 5.2a). The essentially purple/blue tarnish is thought to be the result of the production of CuS on destabilization. The iridescent 'peacock' ore colour may be caused by the presence of a film of an iron oxide: hydroxyl or hydrate complex on a covelline layer (see Figure 5.2b).

Instability is also shown by bornites containing silver,[38] which tends to migrate to the surface to concentrate in the tarnish layer by a diffusion mechanism as yet unknown.

The tarnish on bornite appears to be a self-limiting and non-destructive reaction at low temperatures. Other than by storage in an oxygen-free atmosphere, it is probably not preventable.

Cubanite

No stability data has been recorded but stored specimens of cubanite ($CuFe_2S_3$) soon exhibit a brown or bluish tarnish. Surface oxidation reactions similar to those exhibited by bornite and chalcopyrite are probably responsible.

Blende and wurtite

Wurtite is the unstable rare form of ZnS, which inverts without structural deformation to blende. Blende is extremely stable and does not show any tendency to tarnish or react in air. It is, however, extremely sensitive to the presence of acids, and will rapidly decompose when associated with oxidizing pyrite to form hydrated sulphates such as goslarite.

Greenockite

Greenockite (CdS) is reported to undergo photo-oxidation in moist air to yellow cadmium sulphate[39]. This reaction, however, does not appear to occur in dry air, although museum specimens frequently show a dull tarnish.

Oldhamite

Artificial calcium sulphide is extremely unstable in moist air, oxidizing, via calcium thiosulphate, to calcium sulphate. The mineral oldhamite (CaS) normally occurs coated with sulphate in meteorites, which suggests that it too is unstable in air.

Cinnabar

Cinnabar (HgS) is one of the earliest described sulphide minerals. Cinnabar has for long been known to darken on exposure to light (Plate 5.3). Pliny noted that cinnabar as a pigment was affected by a 'rust' and was damaged by the action of sunlight and even moonlight. Albertus Magnus[40] noted that the oxidation of cinnabar to mercury occurred *in situ* in mines.

Cinnabar will usually tarnish slightly from red to grey on exposure to light and air. The mechanisms are, however, not too clear. Artificial cinnabar darkens rapidly on exposure to light; but it is unlikely that metacinnabarite (HgS) (the black high temperature form) is produced. Under pressure, cinnabar will destabilize to yield metallic mercury.

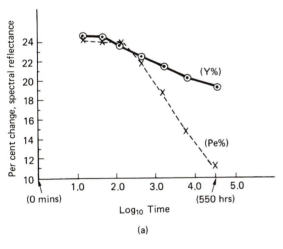

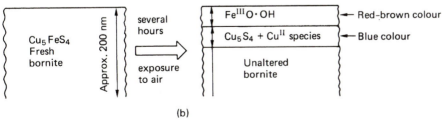

Figure 5.2 (a) Tarnish development on bornite between 10 minutes and 550 hours after exposure to the atmosphere. (b) Schematic diagram showing a possible mechanism for the development of tarnish on bornite (adapted from Vaughan *et al.*, 'The surface properties of bornite', 1987).

Galena

The mineral galena (PbS) is extremely stable under normal storage conditions. Synthetic galena and lead sulphide surfaces are, however, very reactive[41] and micro-crystalline forms tarnish rapidly when exposed to light.[30] Natural galena shows a tendency to tarnish from lead-grey to blue, but little tendency to oxidize at low temperatures. The tarnish may be the result of the presence of a silver sulphide with which galena is almost always associated.

At elevated temperatures oxidation in moist air to sulphates such as anglesite will occur rapidly. In their review of the surface oxidation of galena, Gardner and Woods[41] report on earlier work which demonstrates some oxidation to polythionates and sulphates when lead sulphide is exposed to moist air. Steger and Desjardins[27] demonstrated that at 52°C and high relative humidity galena oxidized, producing thiosulphate. Buckley and Woods[42] showed that at 20°C and 65% RH fresh galena surfaces oxidized slowly, with the surface becoming steadily more enriched with lead; no evidence of lead sulphate was reported even after long exposure to high humidities.

It is probable that appreciable oxidation occurs where galena and pyrite-rich sulphides occur in intimate association, with production of anglesite.

Albandite

Albandite (MnS) is listed as tarnishing brown on exposure to air[14] and oxidizing in moist air, probably to a hydrated manganese sulphate. These reactions have not, however, been well characterized.

Molybdenite

Molybdenite (MoS_2) is not reported as tarnishing but can be oxidized in oxygenated water at low temperature. No stability data are recorded. Avoid storage in high RH environment.

Pyrrhotine

Pyrrhotine is usually iron-deficient ($Fe_{1-x}S$, where ($\times > 0.125$) and very reactive[43]. It rapidly tarnishes from bronze–yellow to dull pink–gold or iridescent on exposure to air. It also oxidizes in moist air to form iron sulphates and iron oxides (Figure 5.3) often with loss of sulphur as hydrogen sulphide (cf the oxidation of pyrite and marcasite, Chapter 6). Sinkankas[19] gives some data on localities which yield particularly oxidation-susceptible specimens.

The oxidation reactions have not been thoroughly investigated, but Steger and Desjardins[44] demonstrated that at 52°C and 68% RH the major oxidation products were goethite and elemental sulphur. Work by Buckley and Woods[45] however, suggests that at 20°C and 65% RH oxidation proceeds rapidly as iron diffuses in the outermost 5–10 nanometres (they demonstrated that oxygen diffuses into the surface to this depth instantaneously upon exposure to air),

Figure 5.3 Crystalline mass of pyrrhotine with developing surface oxidation (whitish areas scattered over whole surface).

and that elemental sulphur does not separate out. This initial reaction occurs in a matter of hours and appears to be self-limiting.

However, massive pyrrhotine specimens undergoing oxidation in collections produce surface growths of iron sulphates. If such growths are not the result of intimate association with reactive phases such as pyrite, the oxidation of pyrrhotine may be relative humidity-dependent. The critical RH may well be near that for pyrite, i.e. near 50%. Steger[44] suggests that at low RH (approx 37%) little oxidation occurs, but that at 50% RH and above diffusion-controlled oxidation of pyrrhotine to iron sulphates and polythionates precedes the formation of goethite.

Mackinawite (FeS), greigite (Fe_3S_4), smythite (Fe_3S_4), mossyite (Fe_xS_y) and kansite (Fe_9S_8), melnikovite (FeS_2)

All these iron sulphides are at best metastable. Mossyite may oxidize within minutes of exposure; greigite and smythite react similarly to pyrrhotine. Mackinawite is very unstable, readily oxidizing to sulphates, iron oxide and hydrogen sulphide when exposed to air.

Violarite and pentlandite

Violarite ($FeNi_2S_4$) tarnishes rapidly from grey to red, but the mechanism of oxidation has not as yet been characterized. Pentlandite ($FeNi)_9S_8$ tarnishes slowly in air.

Bravoite

Bravoite ($Ni,Fe)S_2$ is recorded as oxidizing in moist air by Palache *et al.*,[14] Bannister,[17] and Sinkankas.[19]

The oxidation reactions may parallel those shown by pyrite, although no experimental work appears to have been carried out to corroborate relative humidity dependence.

Millerite

Millerite (NiS) tarnishes rapidly from metallic yellow to iridescent grey. Little information is available on the oxidation of nickel-sulphide minerals. Hydrated nickel sulphates, e.g. morensite and retersite, are, however, commonly associated oxidation products of nickel ores in mines. With both bravoite and millerite, storage at relative humidities above 40% to 50% should be avoided.

Polydymite (Ni_3S_4) and vaesite (NiS_2)

Both these sulphides are reported to slowly tarnish in air. Again, avoid storage in environments where relative humidity exceeds 40%–50%.

Linnaeite (CO_3S_4) and carrollite (($Cu,Co)_2S_4$)

Both these sulphides tarnish rapidly in air from metallic grey to red. Little data on stability is available, but it is likely that limited superficial oxidation to hydrated cobalt sulphates will occur in moist air, and storage below 40% to 50% RH is indicated.

Realgar and Orpiment

Realgar (AsS) undergoes rapid photochemical oxidation to yellow or buff-coloured orpiment and arsenolite (Plate 5.4). The results of the oxidation of realgar have been known from ancient times, but the mechanisms are far from clear. Roberts *et al.*[46] consider that in some cases realgar transforms to para realgar under the influence of light. Daniels[47] described the oxidation products on carved realgar statuettes as a mixture of orpiment and arsenious oxide (arsenolite). The reaction appears to be rapid in sunlight, especially for the more granular varieties. Continued exposure to light will eventually lead to complete disintegration of even well-formed crystals. The oxidation reactions of realgar probably proceed sequentially thus:

$$\underset{\text{realgar}}{AsS} \xrightarrow{\text{Light}} \underset{\text{orpiment}}{AS_2S_3} \xrightarrow{H_2O/O_2} \underset{\text{arsenolite}}{AS_2O_3 + S^0}$$

Side reactions may occur, and form sulphuric acid or sulphur dioxide.

Orpiment (As_2S_3) is not reported as a reactive mineral, but Mellor[30] notes that in moist air synthetic AS_2S_3 will oxidize to arsenious oxide; in oxygenated water at low temperature orpiment oxidizes to arsenious oxide and hydrogen sulphide. Orpiment is not affected by exposure to light.

Realgar should always be stored in light-tight boxes. Specimens for exhibition should be exposed to the lowest possible light levels. Avoid high RH for storage of both realgar and orpiment.

Stibnite

Stibnite (Sb_2S_3) is susceptible to development of dull blue–black tarnish. The reaction appears self-limiting, and is probably photochemically induced. Theoretically crystalline stibnite could transform at low temperature to red amorphous Sb_2S_3, although this is not recorded as occurring in collections. Palache *et al.*[14] record the oxysulphide kermesite (Sb_2S_2O), various oxides and hydrated oxides as oxidation products, but these are rarely produced as oxidation products on specimens in collections. The nature of the blue–black tarnish is unknown. Plate 5.5 shows portions of the same specimen with and without exposure to daylight. The possible reaction sequence is:

$$\underset{\text{steel grey}}{Sb_2S_3} \xrightarrow{\text{light}} \underset{\text{black}}{Sb_2S_3} \xrightarrow{O_2} \underset{\text{blue}}{Sb_2O_3}$$

Bismuthite

Bismuthite (Bi_2S_3) alters readily on exposure to air from metallic lead-grey to a yellow or iridescent tarnish. Surface oxidation to hydrated bismuth oxide is the most likely explanation.

5.2 Sulphosalts: reactivity and care in collections

The sulphosalts as a group include over 100 mineral species with a very wide range of compositions. In essence all contain sulphur with bismuth, antimony, arsenic and sometimes vanadium or germanium. The term sulphosalt originally indicated that the mineral was a 'salt' where sulphur had replaced the oxygen of the anion, e.g. $-AsS_4$ was loosely analogous to $-SO_4$. The anions are, however, far too complex to merit this simple interpretation. They are essentially double sulphides with different crystallographic structure to the 'simple' sulphides. Thus proustite (Ag_3AsS_3) is effectively $3Ag_2S.As_2S_3$.

The properties of these 'double' compounds are, as would be expected, highly variable. Where essentially stable sulphides are present, the mineral may well be inert. Where one or other of the sulphides is unstable, as in the example of proustite, the sulphosalt will tend to be stable and the products of decomposition may comprise two or more phases. Several sulphosalts have been reported as particularly sensitive to light.

The following sulphosalts, according to Pearl[18] and Bannister,[17] exhibit possible light-induced, self-limiting surface changes which may involve oxidation reactions: polybasite, polyargyrodite, pyrostilpnite, xanthoconite, samsonite, diaphorite, freislebenite, miagyrite, aramayoite, matildite, trechmannite, sartonite, vrbaite, fizelyite, jamesonite and hutchinsonite.

All the above except jamesonite, sartorite and vrbaite contain silver. It is therefore most probable that photo-decomposition causes silver to migrate to the surface of the mineral, and that the tarnish is self-limiting, i.e. effectively protects the underlying mineral from further decomposition. It is important to consider, however, that any liberated silver can itself be reduced to an alien paragenesis through the action of atmospheric hydrogen sulphide.

Reactivity and care in collections of sulphosalts

Certain sulphosalts undergo reactions in storage which have been characterized. Descriptions of these follow.

Argyrodite and canfieldite

Argyrodite (Ag_8GeS_6) and canfieldite (Ag_8SnS_6) both tarnish to blue or violet-grey when exposed to light and air.

Epigenite

Epigenite undergoes sequential tarnishing first to black and then blue. It is probable that epignite is a mixture of chalcopyrite, and arsenopyrite and reacts as such. The products of surface oxidation probably include covelline and iron oxides. Avoid exposure to light and high humidity.

Pyragyrite and andorite

Pyragyrite (Ag_3SbS_3) reacts rapidly to intense light,[28] changing from yellow-grey to grey. The reactions involve photo-decomposition to silver and probably stibnite. Pyragyrite specimens in collections often exhibit dull tarnish.

Andorite ($PbAgSb_3S_6$) tarnishes rapidly in light from steel grey to iridescent yellow, possibly via similar reaction to pyragyrite.

Proustite

Proustite ($Ag_3A_5S_3$) reacts rapidly to intense light,[28] changing from transparent red to translucent grey. Exhibited specimens sometimes show this change. See Plate 5.6.

Wittichenite

Palache[14] notes that wittichenite (Cu_3BiS_3) tarnishes from grey-white to pale lead-grey. It will further undergo surface alteration, forming yellow–brown, blue and finally red products, probably a mixture of bismutite and covelline.

Sulvanite

Sylvanite (Cu_3VS_4) tarnishes dull, possibly oxidizing to a copper vanadate (cuprodescloizite). Storage at low relative humidity is suggested, as well as avoiding exposure to light.

Enargite

Enargite (Cu_3AsS_4) undergoes rapid oxidation to copper arsenate under mine conditions. Palache[14] notes that it tarnishes dull. Moritz observes that it will rapidly alter to tennanite at low temperature. Store at low relative humidity (less than 50%). Enargite also appears to readily develop a black velvety efflorescence when exposed to atmospheres containing volatile sulphur compounds.

Jordanite

Jordanite ($Pb_{14}As_7S_{24}$) is recorded as usually tarnished iridescent. Several other lead sulphosalts are known to alter, including bouronite ($PbCuSbS_3$), which darkens on exposure to light; and aikinite ($PbCuBiS_3$), which oxidizes in air possibly to bismutite. Palache[14] records that this mineral tarnishes from grey to brown or copper-red, sometimes with a yellowish coating. All lead sulphosalts should be protected from light and stored at low relative humidity.

Boulangerite

Boulangerite (Pb_5SbS_{11}), besides the silver sulphosalts, is probably the most reactive in the group. Palache[14] notes that the blue–grey mineral is often covered with yellow spots, probably of bindheimite ($Pb_2Sb_2S_6(O,OH)$). Oxidation reactions probably occur at low relative humidity.

Smithite

Smithite ($AgAsS_2$) alters photochemically from red to orange–red, possibly to orpiment. Avoid exposure to light.

Lorandite

Lorandite ($T1AsS_2$) is a carmine-red mineral which will photo-decompose superficially, forming a powdery translucent-orange or yellow surface coating, possibly a mixture of orpiment and thallium compounds together with sulphur dioxide[48]. Avoid exposure to light.

Chalcostibnite

Chalcostibnite ($CuSbS_2$) tarnishes from grey metallic to dull blue, probably due to photo-oxidation. Avoid exposure to light.

Benjamite

Benjamite ($Pb(Cu,Ag)Bi_2S_4$) is reported as tarnishing from grey metallic to dull yellow or red, probably because of the formation of bismuthite.

Plate 2.1 *Brown topaz from Utah, USA, fades in light; on heating for 1 hour to the temperature shown, the colour is lost well below 200°C.*

Plate 2.2 *Matched red tourmalines heated for 1 hour to 200°C, 250°C, 300°C, and 400°C. Since this colour centre is stable at 200°C, it does not fade in light.*

Plate 2.3 *Deep blue Maxixe-type irradiated beryl (left); natural Maxixe beryl partly faded (right) and fully faded (centre) by exposure to light.*

Plate 2.4 *Normal-appearing faceted emerald, consisting of whitish cracked beryl dyed with green and yellow dyes, viewed in diffuse transmitted light. These are organic dyes that will fade in light. Photograph courtesy Gemological Institute of America; copyright* Gems and Gemology, *used with permission.*

Plate 3.1 *Coquimbite crystals, Emery Co., Utah, each crystal ~1 cm across:*

(a) Showing the lavender colour and vitreous lustre typical of fresh, unaltered specimens. NMNS #40986.

(b) Showing alteration of the crystal surface to iron oxyhydroxides as a result of hydrolysis during periods of deliquescence. NMNS #38551.

Plate 3.2 *Fibrous growths of bilinite on a formerly equant crystal of melanterite. A result of storage at mineral–solution 'equilibrium' conditions.*

Plate 3.3 *A meta-autunite crystal formed by the dehydration of an autunite crystal. This has not resulted in decrepitation but has caused opening along the [001] cleavage. NMNS #38206. Width of specimen as viewed is 3cm.*

Plate 3.4 *A group of synthetic chalcanthite crystals, partly effloresced shows the lack of cohesion in the dehydration product, evidenced by cracks in the pale blue bonattite. Width of specimen as viewed is 8cm.*

Plate 4.1 *Extensive development of a hydrated calcium chloride nitrate/acetate triple salt on a limestone cast of a Jurassic ammonite from Chile, stored in a case constructed partly of oak (courtesy of The Natural History Museum, London).*

Plate 5.3 *Cinnabar exhibiting partial darkening after exposure to light.*

Plate 5.1 *Iridescent tarnish on surface of chalcopyrite intergrowths with quartz; the bright silver–gold colour of freshly fractured chalcopyrite can be seen centre left (courtesy of the Natural History Museum, London).*

Plate 5.4 *Realgar aggregate showing (a) light degradation to yellow products on exposed front of specimen.*

(b) Kept away from direct light unchanged (courtesy of The Natural History Museum, London).

Plate 5.2 *Bornite showing typical peacock blue tarnish, which often develops within minutes of exposure to air (courtesy of The Natural History Museum, London).*

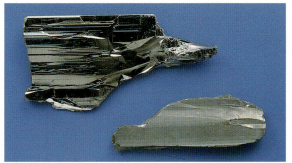

Plate 5.5 *Two fragments taken from the same stibnite specimen. The fragment bottom right was exposed under daylight conditions for several months, the other fragment was stored in the dark (courtesy of The Natural History Museum, London).*

Plate 5.7 *Nickeline-rich shale sample, showing extensive development of oxidation products (courtesy of The Natural History Museum, London).*

Plate 5.6 *Samples of proustite (a) stored in dark (courtesy of The Natural History Museum, London).*

(b) Exposed to light (courtesy of M. Price, University Museum, Oxford).

Plate 6.1 *Fine pyrite specimen, Brazil (courtesy of The Natural History Museum, London).*

Jamesonite and frankcëite

Jamesonite ($Pb_4FeSb_6S_{14}$) and frankcëite ($Pb_5Sn_3Sb_2S_{11}$) are sometimes reported as tarnishing from grey metallic to iridescent.

Rathite and baumhaurite

Rathite ($Pb_{13}As_{18}S_{49}$) and baumhaurite ($Pb_4As_6S_{13}$) are both reported as developing an iridescent tarnish.

Semseyite and fuloppite

Semseyite ($Pb_9Sb_8S_{21}$) and fuloppite ($Pb_3Sb_8S_{15}$) are both said to tarnish from lead-grey to dull and steel-blue/bronze respectively.

Galenobismutite

Galenobismutite ($PbBi_2S_4$) may undergo slow oxidation to bismuthite.

Zinkenite, berthierite and marrite

Zinkenite ($Pb_6Sb_{14}S_{27}$), berthierite ($FeSb_2S_4$) and marrite (unknown chemistry) all develop iridescent tarnish.

5.3 Antimonides, arsenides, bismuthides, selenides, sulphantimonides, sulpharsenides and tellurides

Together with the sulphides and sulphosalts, the antimonides, arsenides, bismuthides, selenides, sulpharsenides and tellurides (henceforth referred to here as the antimonides, etc.) form a large and very important class of minerals, including as it does a number of ores of commercial importance. The enormous diversity of compounds comprising the class exhibit differing physical properties and chemical behaviour with even closely similar minerals, in terms of elemental composition, showing substantial differences in susceptibility to environmental influences.

There are at present some 120 known antimonides, etc. Of this number at least 25 are recorded as unstable under the range of conditions found in collections. Changes reported range from slight tarnishing under the influence of light and/or oxygen, to extensive surface alteration, sometimes accompanied by structural damage brought about by oxidation to hydrated products after exposure to moist air.

The following list of minerals, together with stability data is not intended to be exhaustive. Few investigations have been carried out into low temperature transformations in this group.

Berzelianite

When freshly collected, berzelianite (Cu_2Se) is silver white with metallic lustre. Palache *et al.*[14] noted that tarnishing occurred very rapidly in air. Bannister[17]

suggested that the reaction appeared to be induced by exposure to light and air. Berzelianite usually contains some silver, possibly an eucairite ($CuAgSe$), and the tarnish to grey may be caused by the electronic destabilization of this mineral and migration of metallic silver to the surface.

Umangite and klockmannite

Umangite (Cu_3Se_2) rapidly tarnishes from dark red with metallic lustre to dull violet-blue. It is likely that limited surface alteration to blue chalcomenite ($Cu_4(SeO_3)_4.8H_2O$) occurs at high relative humidity. Klockmannite ($CuSe$) exhibits a similar reaction, tarnishing from grey to blue. For storage and exhibition, avoid exposure to relative humidity in excess of 50%.

Altaite and breithauptite

Altaite ($PbTe$) tarnishes from black to dull brown, and breithauptite ($NiSb$) tarnishes from copper red to violet, upon exposure to air.

Rickardite and weissite

Both these copper telluride minerals tarnish, respectively from purple and dark blue to black.

Krennerite ($AuTe_2$), calaverite ($AuTe_2$) and Sylvanite ($Ag\,AuTe_4$)

Each of these tellurides is reported as 'tarnishing' from silver-white to brass-yellow, probably due to diffusion and exsolution of gold. All are reported by Palache *et al.*[14] as altering to gold structurally. Sylvanite also alters to tellurite. Except in the case of sylvanite, whether or not these effects are photo-induced is not known with any degree of certainty.

Tetradymite

Tetradymite (Bi_2Te_2S) is pale grey when fresh but rapidly tarnishes dull or iridescent. Surface alteration to bismuthite and the hydrated tellurite montanite probably occurs in moist air.

Lollingite

Lollingite ($FeAs_2$) is, under humid conditions, likely to oxidize slowly to the hydrated ferrous and ferric arsenates scorodite and symplesite; both are reported as alteration products in mineworkings. Walker,[48] however, observed that lollingite showed no sign of surface oxidation when stored in moist air under a bell-jar for one week.

Arsenopyrite

Arsenopyrite ($FeAsS$) is generally considered to be relatively stable under normal storage conditions. Exposure to high relative humidity and temperatures in excess of $30°C$ will, however, cause some oxidation of the mineral to scorodite ($FeAs_5O_4.2H_2O$) and possibly hydrated iron sulphates. Specimens often

Figure 5.4 Radiating arsenopyrite, showing surface growths of black oxidation product. (×10)

Figure 5.5 'Stress' cracking is sometimes seen in arsenopyrite specimens and other radiating and nodular sulphides. (×0.75)

show a dull brown or grey tarnish, and occasionally grey/black crystalline growths are seen to develop slowly on fractures, especially of radiating or prismatic varieties, as shown in Figure 5.4. Granular and massive specimens sometimes also show signs of incipient stress cracking (see Figure 5.5). Richardson and Vaughan[53] suggested that arsenopyrite oxidizes direct to sulphates in air, but no mechanisms are indicated.

Safflorite
Safflorite ($CoFeAs_2$) rapidly tarnishes from silver-white to dark grey. Walker[48] reported experimental oxidation to arsenate under aqueous conditions; it is therefore likely that exposure to humid air would accelerate its oxidation to erythrite. Store at less than 50% RH.

Smaltite and chloanthite
Smaltite ($CoNiAs_3$) and chloanthite ($(Ni,Co)As_3^{-x}$) both occasionally develop slight tarnish. Both have been reported to oxidize in a similar fashion to pyrite, with formation of nickel and cobalt arsenates.[14] Walker[48] noted, however, that there were no

signs of oxidation after a sample of the smaltite was stored at high humidity for a week. Sinkankas[19] observed that some specimens of chloanthite were unstable and oxidized in damp air. Storage at low relative humidity is recommended.

Cobaltite and Glaucodot
Cobaltite ($CoAsS$) and glaucodot ($(CoFe)AsS$) both tarnish.[14,16] Cobaltite almost always rapidly develops a pink–blue tarnish. This tarnish may in fact be a coating of cobalt sulphates and arsenates. Buckley, however, has shown that the experimental oxidation of cobaltite surfaces[49] in air results in the rapid formation of arsenious oxide together with the slower formation of cobaltous oxide. The sulphur phase is not thought to be involved, at least initially, in these reactions. Glaucodot occasionally develops a brown tarnish, possibly of an iron oxide.

Penroseite and melonite
Penroseite ($NiCuSe_2$) and melonite ($NiTe_2$) will tarnish. Melonite tarnishes white to brown, while penroseite dulls.

Nickeline

Nickeline (Niccolite) (NiAs) tarnishes rapidly from bright copper-red to grey or black and dull, especially in humid air. Parsons[16] showed that nickeline was oxidized to annabergite (NiAsO) by exposure to moist air. Plate 5.7 shows a sample of an oxidized nickeline-rich shale which had been stored in the 40%–75% relative humidity range for an unknown period of time. Storage and exhibition at low relative humidity (30% or less) is recommended.

Rammelsbergite and pararammelsbergite

Rammelsbergite ($NiAs_2$) and pararammelsbergite ($NiAs_2$) will oxidize extremely rapidly to annabergite in moist air under humid storage conditions. Walker[49] observed that rammelsbergite oxidized extensively within a week to annabergite when stored in a bell-jar over a small quantity of water. Palache *et al.*[14] mistakenly cites erythrite (CoAsO) as an oxidation product of pararammelsbergite. Storage and exhibition at low relative humidity (30% or less) are recommended.

Gersdorffite

Gersdorffite (NiAsS) tarnishes from metallic silver-white to grey or grey-black. Parsons[16] and Bannister[17] cite gersdorffite as showing similar oxidation reactions to pyrite and rammelsbergite, although these have not been investigated. Treat as humidity-sensitive.

Petzite

Petzite (Ag_3AuTe_3) tarnishes rapidly from steel-grey or black to yellow–gold, probably brought about by photo-induced migration of gold to the surface.[28] Avoid exposure to sunlight and UV sources.

Hessite

Hessite (Ag_2Te) is recorded as light sensitive by Bannister,[19] Stephens[28] and Criddle *et al.*[50] Exhibited specimens often show dulling of metallic lustre with development of yellow or grey tarnish.

On exposure to intense light, hessite is reported as developing a pale yellow film which spreads rapidly over exposed surfaces.[28] Prolonged exposure causes surface rearrangement, and Criddle *et al.*[50] speculate that tellurium migrates out of the hessite into adjacent phases (usually minerals such as acanthite, see Figure 5.6).

Joseite

Joseite (Bi_3TeS) tarnishes from steel-grey to dull grey or iridescent. Palache *et al*[14] note that alteration to montanite ($BiTeO_6.2H_2O$) occurs, suggesting susceptibility to moist air.

Wehrlite

Wehrlite (Bi_3Te_2) tarnishes from bright metallic to dull.

Domeykite and algonodite

Domeykite (Cu_3As) and algonodite (Cu_6As) are both reported as tarnishing and becoming dull on exposure to air and possibly light. Domeykite undergoes extensive sequential alteration, from white-grey metallic to yellow to brown tarnish and finally iridescent. The brown coating has not been characterized but may contain arsenolite.

Dyscrasite

Dyscrasite (AgSb) usually tarnishes from silver-white to dull grey to yellow or black, probably the result of conversion to antimony oxide.

Cervelleite

The photochemical reactions which cervelleite (Ag_4TeS) undergo are described in detail by Criddle *et al.*[50] This new mineral is found in association with acanthite and hessite. The three minerals undergo what is probably mutual surface migration or diffusion of silver, tellurium and sulphur between the phases. Figure 5.6 shows a schematic representation of the photochemical alterations occurring in a freshly polished surface containing acanthite, cervelleite and hessite. Similar reactions are reported by Jeppson[51] on synthetic chalcopyrite, isocubanite and pyrrhotine in contact with jalpaite-acanthite. Chen *et al.*[34] and Remond *et al.*[52] report silver diffusion on the surface of chalcopyrite and tennanite.

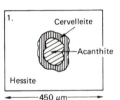

1. Freshly polished surface showing acanthite/cervelleite inclusion in hessite.
←—— 450 µm ——→

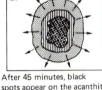

2. After 45 minutes, black spots appear on the acanthite, the cervelleite's reflectance drops and an alteration zone starts to affect the hessite.

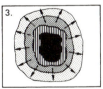

3. After 90 minutes the acanthite is almost black, the cervelleite is unchanged but the hessite continues to alter.

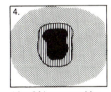

4. After 24 hours acanthite shows no further change, the cervelleite looks like unaltered acanthite and the hessite becomes darkened up to twice the diameter of the acanthite/cervelleite inclusion.

Figure 5.6 Schematic diagram depicting the photochemical alterations undergone by acanthite, cervelleite and hessite intergrowth (adapted from Criddle *et al.*, 1989).

References

1 KOSTOV, I. and MINCEVA-STEFANORA, J., *Sulphide Chemistry, Crystal Chemistry, Paragenesis and Systematics*, Bulgarian Academy of Sciences (1982)

2 DANA, J.D. and DAND, E.S., *A System of Mineralogy*, Wiley, New York and London (1892)

3 YARAR, B. and SPOTTISWOOD, D.J. (Eds), 'Interfacial Phenomena and Mineral Processing, *Proceedings of the Engineering Foundation Conference*, Engineering Foundation, New York (1982)

4 VAUGHAN, D.J. and CRAIG, J.R., *Mineral Chemistry of Metal Sulphides*, Cambridge University Press, Cambridge (1978)

5 HOLMYARD, E.J., *Alchemy*, Penguin, Harmondsworth (1957)

6 BRION, D., 'Étude par spectroscopie de photoelectrons de la dégradation superficialle de FeS$_2$, CuFeS$_2$, ZnS et PbS à l'air dans l'eau', *Application of Surface Science*, 5, pp. 133-52 (1980)

7 BUCKLEY, A. and WOODS, R., 'Investigation of the surface oxidation of sulphide minerals via ESCA and electrochemical techniques', in 'Interfacial Phenomena in Mineral Processing', Eds Tarrar, B. and Spottiswood, D.J., *J. Engineering Foundation*, New York (1983)

8 BUCKLEY, A.N., HAMILTON, I.C. and WOODS, R., 'Studies of the surface oxidation of pyrite and pyrrhotite using x-ray photoelectron spectroscopy and linear sweep voltammetry', *Proceedings of the Electrochemical Society*, 19, pp. 88-121 (1988)

9 BEIGLER, T. and HORNE, M.D., 'The electrochemistry of surface oxidation of chalcopyrite', *Proceedings of the Electrochemical Society*, 84-10, pp. 321-39 (1984)

10 VAUGHAN, D.J., TOSSELL, J.A. and STANLEY, G.J., 'The surface properties of bornite', *Mineralogical Magazine*, 51, pp. 285-93 (1987)

11 BURKIN, A.R., *The Chemistry of Hydrometallurgical Processes*, E & F.N. Spon, London (1966)

12 SMITH, E.E., 'Engineering aspects of acid mine drainage', *Proceedings, Second Annual Symposium Water Resources Research*, Ohio State Chemistry (1966)

13 BRUYNESTEYN, A. and HACKEL, R.P., 'Evaluation of acid production potential of mining waste materials', *Minerals and the Environment*, 4, pp. 4-9 (1982)

14 PALACHE, C., BERMAN, H. and FRONDEL, G., *The System of Mineralogy of J.D. & E.S. Dana*, 7th edition, Vols 1 and 2, Wiley, New York (1944, 1956)

15 PEARL, R.M., *Mineral Collectors Handbook*, Mineral Book Company, Colorado Springs (1948-9)

16 PARSONS, A.L., 'The preservation of mineral specimens', *American Mineralogist*, 7, pp. 59-63 (1922)

17 BANNISTER, F.A., 'The preservation of minerals and meteorites', *Museums Journal*, 36, pp. 465-76 (1937)

18 PEARL, R.M., *Cleaning and Preserving Minerals*, Earth Science Publishing Company, Colorado Springs (1975)

19 SINKANKAS, J., *Gemstones and Mineral Data Book*, Winchester Press, New York (1972)

20 HOWIE, F.M.P., 'Conservation and Storage: Geological Material', in *Manual of Curatorship*, Editor Thompson, M.A. Butterworths, London & Boston, pp. 308-22 (1984)

21 NASSAU, K. (see Chapter 2, this volume)

22 HENDERSON, B., *The Structures and Properties of Solids 1: Defects in Crystalline Solids*, Edward Arnold, Bath (1972)

23 SHUEY, R.T., *Semiconducting Ore Minerals*, Elsevier, New York (1975)

24 MITCHELL, D. and WOODS, R., 'Analysis of oxidized layers on pyrite surfaces by x-ray emission spectroscopy and cyclic voltammetry', *Australian Journal of Chemistry*, 31, pp. 27-34 (1978)

25 HOWIE, F.M.P., 'Pyrite and Conservation', *Newsletter of the Geological Curators Group*, 1, pp. 457-65 (1977)

26 BOULEAQUE, J., 'Simultaneous determination of polysulphides and thiosulphates as an aid to ore exploration', *Journal of Geochemical Exploration*, 15, pp. 21-36 (1981)

27 STEGER, H.F. and DESJARDINS, L.E., 'Oxidation of sulphide minerals: III determination of sulphate and thiosulphate in oxidized sulphide minerals', *Talanta*, 24, pp. 455-60 (1977)

28 STEPHENS, M.M., 'Effects of light on polished surfaces of silver minerals', *American Mineralogist*, 16, pp. 532-49 (1931)

29 GUILD, F.N., 'A microscopic study of silver ores and associated minerals', *Economic Geologist*, 12, pp. 297-353 (1917)

30 MELLOR, J.W., *Treatise on Organic Chemistry*, Longman, London (1933)

31 WHITEHEAD, W.L., 'Technique of Mineralography', *Economic Geologist*, 12, pp. 697-716 (1917)

32 ALLEN, R.L. and MOORE, W.J., 'Diffusion of silver in silver sulphide', *Journal of Physical Chemistry*, 63, pp. 223-6

33 BUCKLEY, A.N. and WOODS, R., 'An x-ray photoelectron spectroscopic study of the oxidation of chalcopyrite'. *Australian Journal of Chemistry*, 37, pp. 2403-13 (1984)

34 CHEN, T.T., DUTRIZAC, J.E., OWENS, D.R. and LAFLAMME, J.H.G., 'Accelerated tarnishing of some chalcopyrite and ternantite specimens', *Canadian Mineralogist*, 18, pp. 173-80 (1980)

35 RAGUSA, S. and MADGWICK, J.C., 'Bacterial leaching of chalcopyrite at low relative humidities', *Proceedings of the Australasian Institution of Mining and Metallurgy*, 289, pp. 229-31 (1984)

36 THORPE, R.I., PRINGLE, G.J. and PLANT, A.G., 'Occurrence of selenide and sulphide minerals in bornite ore of Kidd Creek massive sulphide deposit, Timmins, Ontario', *Geological Survey of Canada Papers*, 76-1A, pp. 311-17 (1976)

37 CHEN, T.T. and PETRUK, W., 'Electron Microprobe analyses of silver bearing minerals in samples collected from the Heath Steel mill, New Brunswick in March 1977', *CANMET Reprint MRP/MSL 78-23 (1R)* (1978)

38 BUCKLEY, A.N. and WOODS, R., 'An x-ray photoelectron spectroscopic investigation of the tarnishing of bornite', *Australian Journal of Chemistry*, 36, pp. 1793-1804 (1983)

39 LICHENSTEIGER, M. and WEBB, C., 'Photon-hole induced adsorption of oxygen on CdS', *Surface Science*, 154, pp. 455-64 (1985)

40 ALBERTUS MAGNUS, *Book of Minerals,* translated by D. Wyckoff, Clarendon Press, Oxford (1967)

41 GARDNER, J.R. and WOODS, R., 'A study of the surface oxidation of galena using cyclic voltammetry', *Journal of Electroanalytical Chemistry*, 100, pp. 447-59 (1979)

42 BUCKLEY, A.N. and WOODS, R., 'An x-ray photo-electron study of the oxidation of galena', *Application of Surface Science*, 17, pp. 401-44 (1984)

43 MUKHERJEE, A.D. and SEN, R., 'Pyrrhotite alteration–a study on Norwegian and Indian sulphide ores', *Geological Society of India Journal*, 23, pp. 196-8

44 STEGER, H.F., 'Oxidation of sulphide minerals VII: effect of temperature and relative humidity on the oxidation of pyrrhotite', *Chemical Geology*, 35, pp. 281-95 (1982)

45 BUCKLEY, A.N. and WOODS, R., 'X-ray photoelectron spectroscopy of oxidized pyrrhotite surfaces: I exposure to air', *Applications of Surface Science*, 22, pp. 280-7 (1985)

46 ROBERTS, A.C., ANSELL, H.G. and BONARDI, M., 'Pararealgar, a new polymorph of AsS from British Columbia', *Canadian Mineralogist*, 18, pp. 525-7 (1980)

47 DANIELS, V., 'Chinese realgar figurines: a study of their deterioration and method of manufacture', *MASCA Journal*, 2, pp. 170-2 (1983)

48 WALKER, T.L., 'Arsenides from the silver veins of South Lorrain, Ontario', *University of Ontario Studies Geological Series No 20: Contributions to Canadian Mineralogy*, pp. 49-53 (1925)

49 BUCKLEY, A.N., 'The surface oxidation of cobaltite, *Australian Journal of Chemistry*, 40, pp. 231-9 (1987)

50 CRIDDLE, A.J., CHISHOLM, J.E. and STANLEY, C.J., 'Cervelleite, Ag_4TeS, a new mineral from the Bambolla Mine, Mexico, and a description of a photo-chemical reaction involving cervelleite, acanthite and hessite', *European Journal of Mineralogy*, 1, pp. 371-80 (1989)

51 JEPPSON, M., 'Photochemically accelerated surface-diffusion of silver on chalcopyrite, isoanbonite and pyrrhotite', *Canadian Mineralogist*, 26, pp. 973-8 (1988)

52 REMOND, G., HOLLOWAY, P.H., HOYLAND, G.T. and OLSON, R.R., 'Bulk and surface silver diffusion related to tarnishing of sulphides', *Scanning Electron Microscopy*, 111, pp. 995-1011 (1982)

53 RICHARDSON, S. and VAUGHAN, D.J., 'Arsenopyrite: a spectroscopic investigation of altered surface', *Mineralogical Magazine*, 53, pp. 223-9 (1989)

6

Pyrite and marcasite

Frank M. Howie

The iron disulphide minerals include ferroselite, melnikovite, marcasite and pyrite.[1] The first two minerals are rare sulphides and are described in Chapter 5. This chapter deals with marcasite and more especially pyrite, because they undoubtedly contribute to the majority of conservation problems found in geological collections. Pyrite is one of the commonest sulphides, occurring across the geological spectrum; marcasite occurs less frequently but causes similar problems.

Pyrite is one of the most widespread and persistent of minerals, and occurs in sediments from Precambrian to Recent. It also occurs in most types of igneous and metamorphic rocks, in the hydrothermal zones around altered ore deposits and volcanic fumeroles, and as a major constituent in sulphide ore deposits. Pyrite normally occurs in distinctively crystalline forms, such as cubic, octahederal or pyritohedral; interpenetrant growths are common, as are striated faces (especially of cubes). Pyrite also occurs massively in nodules, finely disseminated in sediments, replacing or infilling fossils and archaeological objects, and as aggregates of minute crystals, including the ubiquitous framboidal pyrite (Figure 6.1).

In all the above forms, whatever the size of the pyrite crystal (several centimetres in diameter for well-grown cubes down to 0.1 to 1.0 micrometres for the individual microcrystals constituting framboids), it is often euhedrally formed. Even in its dendritic form, which is found in nodules and some fossil infillings, poorly formed, low index-faces can generally be ascertained.

Chemically pyrite is FeS_2, with a very limited stoichiometric range. However, it often contains trace quantities of elements such as cobalt, arsenic, gold, nickel, copper, etc. Pyrite is the cubic dimorph of FeS_2. Its symmetry is hemihedral with the [100] face often being the most developed or strongest

growing. Physical properties such as conductivity, thermo-EMF and adsorptivity vary between the faces of pyrite, as do surface micropictographic details such as etch pits, growth layers and dislocations.

Marcasite is a very much rarer mineral than pyrite with a range of occurrence limited principally to sedimentary rocks, especially clays, limestones in alteration zones, and in sulphide ore bodies. It commonly occurs as tabular crystals, often repeatedly twinned, forming low bipyramids, flat radiating aggregates, platelets and massive forms. Chemically marcasite is FeS_2, with limited stoichiometric range; trace impurities are probably rarer than in pyrite. Marcasite is the orthorhombic dimorph of FeS_2. Distinction between pyrite and marcasite, where doubtful, is ascertained by XRD or reflectance microscopy. Of the two dimorphs, marcasite is generally cited as the least stable in air, but many forms of pyrite are probably equally as unstable.

Common synonyms for pyrite include pyrites, iron pyrites, fool's gold, mundic and, in part, marcasite. The last two are little used nowadays. Synonyms for marcasite include white pyrites, radiated pyrites, hydropyrite and cockscomb pyrites, none of which are in current use. As is explained in Section 6.2, there has been considerable confusion between the two minerals in terms of identification, occurrence and reactivity.

6.1 Oxidation and conservation

The deterioration and loss of mineral and rock specimens caused by the oxidation of pyrite and marcasite are certainly two of the most serious conservation problems encountered in collections of geological material.[2,3,4,5] Indeed they are increasingly recognized to be problems in collections of archaeological material and antiquities,[6] and in products

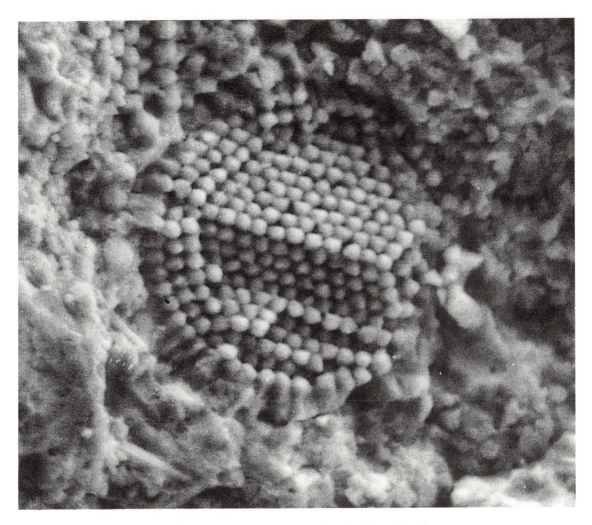

Figure 6.1 Framboidal pyrite from pyritic nodule, Eocene, UK. (SEM, Scale bar 10 pm.)

such as concrete[7] and building foundations,[8] as well as in the storage of reference sulphide ores.[9]

The oxidation of these two minerals in geological collections has attracted attention for centuries, and has become commonly, but perhaps misleadingly, known as pyrite 'disease', pyrite 'rot' or, more simply, pyrite decay. Many proposals have been put forward to explain the oxidation, and numerous treatments have been proposed and tried. Few have been demonstrably successful, however, and large numbers of valuable mineral specimens, assemblages and rock samples have been destroyed in collections throughout the world.

By no means all pyrite and marcasite is susceptible to oxidation. Well-formed crystals of pyrite from several localities retain their brilliance indefinitely, and some, such as some from Elba and Brazil, are among the most spectacular and stable of minerals (see Plate 6.1). On the other hand, the pyrite commonly found in mineral assemblages and fossil material,[3,10] as well as the nodular varieties of both pyrite and marcasite, are often highly susceptible to oxidation.

The chief products of the oxidation process are sulphuric acid and various hydrated sulphates, which are often seen as lightly coloured crystalline growths on the surface of affected material (Figure 6.2). The surface alterations observed are simply the outward signs of oxidation. The phase change from sulphide to sulphate causes a very large volume increase, which leads to internal stress and bulk destruction of specimens. The acidic oxidation products, sometimes seen as a wet glistening film on the specimen's surface will, given the opportunity,

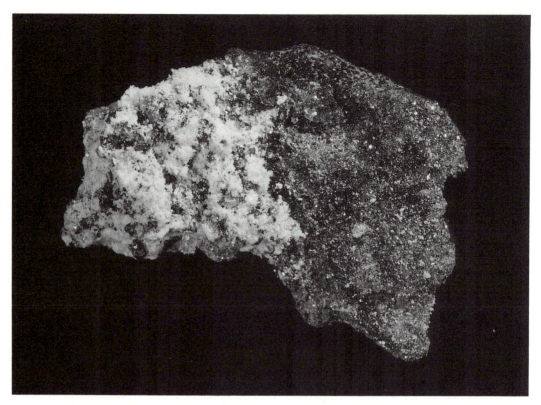

Figure 6.2 Oxidized pyritic specimen showing typical growth of hydrated sulphates and resultant distortion. (×1)

destroy labels, boxes and, in extreme cases, even wooden storage cabinets. In mineral assemblages other minerals associated with oxidizing pyrite and marcasite are often destroyed, with the formation of secondary minerals such as gypsum, jarosite, vivianite and anglesite. The extent of deterioration depends on the type and amount of pyrite or marcasite undergoing oxidation, and the prevailing storage environment. See Bannister,[2] Pearl,[5] Howie[3,10] and Lodha *et al.*[11] for detailed discussions on this aspect.

There is no smooth transition between the stable and oxidation-susceptible types of pyrite and marcasite. The term reactive or unstable pyrite[3,12] is, however, generally given to material which, because of its microcrystalline nature, is highly susceptible to oxidation. The term stable or inert pyrite is usually reserved for the well-crystallized varieties. However, with some well-crystallized pyrites, and some marcasites, brown or orange-coloured surface tarnish, or, in some cases, a whiteish bloom, may develop. The bloom is probably caused by self-limiting growth of iron sulphates on reactive portions of otherwise stable material. The tarnishing of otherwise bright, well-crystallized pyrite is more difficult to explain, but may be the result of diffusion controlled oxida-

tion to iron oxides such as limonite or iron–sulphur–oxygen species on a molecular scale.[13]

Efforts to find simple and effective methods for the prevention and treatment of deterioration have to a great extent been hindered because, in spite of excellent technical reviews by Radley[14] and Bannister[15] in the 1920s and 1930s, the oxidation mechanisms for pyrite and marcasite in collections remained largely misunderstood for several decades.

Two major hypotheses prevailed up to the late 1970s. The oldest idea maintained that purely chemical mechanisms were responsible, and that oxidation (and therefore deterioration) could be prevented simply by removing oxidation products and excluding air by coating with lacquers or resins, or by immersion in inert liquids. During the 1960s and 1970s the theory that certain types of bacteria (thiobacteria) were responsible for museum specimen decay took hold, especially in the UK. This idea was attractive in that, if correct, oxidation could simply be prevented by the use of bactericides without exclusion of air. This mistaken belief unfortunately led to the loss and deterioration of enormous numbers of mineral and fossil specimens in Europe and elsewhere.[10]

During the 1970s and 1980s research into oxidation mechanisms for pyrite under a variety of experimental conditions clearly demonstrated that the problem was far more complex than had hitherto been thought. Application of some of these findings to the conservation of pyrite and marcasite in collections is currently producing encouraging results. Before considering modern pyrite oxidation theory, it is instructive reviewing the historical aspects.

6.2 Historical review of oxidation

For several hundred years 'pyrites' was the general term for all brassy yellow or white sulphides, including chalcopyrite, true pyrite, arsenopyrite, pyrrhotine, and marcasite. The word marcasite (as marchasite) is probably of Spanish or Arabic origin, and was used from medieval times up to the end of the eighteenth century to describe well-crystallized 'pyrites'. During the same period nodular varieties were termed 'pyrites' and the softer white or brown varieties were termed 'wasserkies'. This last group probably included true marcasite, pyrrhotine and some forms of pyrite.

Dana[16] records 'wasserkies' as an early translation of *pyrites aquosus*, perhaps alluding to tendencies for some specimens to become wet when undergoing deterioration. Wallerius,[17] however, includes *pyrites aquosus* with 'vatenkies' and *pyrites fuscus* as terms applied to white, arsenic-rich varieties of pyrite. By the late 1700s the term *pyrites aquosus* seems to have disappeared.[18,19] 'Wasserkies' may also be a corruption of 'weisserkies' (white iron pyrites, e.g. marcasite).

It seems very likely that this nomenclatural confusion led to the term marcasite being used extensively during the late eighteenth and nineteenth centuries to describe most if not all easily oxidizable iron disulphides, and this connotation still lingers. Davies,[20] Shepherd[21] and Sinkankas[22] all denote fossils and nodules as 'marcasitic', because they decay in air, when in fact they contain only pyrite.

Probably the earliest reference to the effects of pyrite oxidation is contained in the introductory passages to *De Lapidibus*[23] (third century BC). Reference is made to a stone found in mines in Thrace which burned when it was split and placed in the sun. This stone was called *spinos* by Aristotle, and was probably a shaly pyritic lignite which reacted vigorously when sprinkled with water. *Spinos* on translation from the Greek is also Theophrastus's name for pyrite.[16]

Pliny and others refer to the methods for preparing vitriols (melanterite and chalcanthite) and misy (copiapite) by weathering copper and iron sulphide ores, and for centuries these processes were widely used to prepare copper and iron sulphates and sulphuric acid. Pliny[24] also referred to a blackish mineral, which he called 'pyrites', that burned the hand when it was rubbed. This possibly referred to the irritant effect on the skin of acidic pyrite oxidation products.

Although both Pliny and, later, Albertus Magnus, made many observations on the stability of minerals and gems, little appeared in the literature concerning direct observation of the oxidation of pyrite until Agricola[25] described in detail the encrustations found on oxidized 'pyrite' nodules.

The first detailed empirical observations were made by Mayow,[26] who noted that 'marcasite' formed vitriol by the combination of 'nitro-aerial spirit', i.e. oxygen, with the sulphur portion of the ore, forming acid substances which in turn combined with the iron portion. Boyle[27] and Lavoisier[28] observed that pyrite and marcasite tended to increase in weight as they vitriolized in air.

Henckel[29] observed that moisture in air was a factor in vitriolization, and that iron disulphides varied in their susceptibility towards oxidation, with nodular and radiating 'pyrites' being less stable than crystalline and laminated varieties. He postulated that the less stable forms were of lower density, were more granular, and were richer in copper or arsenic than the more stable forms. To explain instability Werner[30] cited arsenic contamination, and Berzelius[31] suggested manganese enrichment. Observations on the progress of oxidation were rare, but Hausmann[32] noted that certain iron pyrites deteriorated by first tarnishing, followed by surface 'rusting', and finally vitriolization.

The earliest chemical analyses of iron sulphides, by Hatchett,[33] refuted an earlier view[34] that sulphur deficiency in pyrite was the prime cause of instability. Berzelius[31] analysed various iron sulphides and pyrite oxidation products, and concluded that unstable pyrites were mixtures of stable iron disulphide (the crystalline portion) and unstable iron monosulphide cement. He proposed that the oxidation proceeded electrochemically, and that cells were set up between the two sulphides when exposed to damp air. However, later analyses[35] showed that pyrites and marcasite contained only FeS_2

Later theories suggested that state of aggregation was responsible for instability[36] and that well-formed stable pyrite crystals had smooth faces, offering few possible sites for the attack of oxidizing agents.[37] Kimball[38] suggested that marcasite (probably unstable pyrite) decomposed more readily than pyrite, because the former occurred in several states of crystallization and aggregation favourable to oxidation, whereas the latter occurred mainly as a compact mineral. Julien later concluded that many specimens of unstable pyrite were not pure, but were mixtures of pyrite and marcasite,[39] with the most unstable specimens containing the highest proportion of marcasite with granular texture.[40] He stated, however, that large, well-formed crystals of marcasite were as stable as similarly formed crystals of pyrite, but that if these stable crystals were crushed to powder, oxidation rapidly ensued after exposure to air. He further suggested that the amount of moisture in the air was probably a critical factor.

Improved analytical and observational techniques have not substantiated many of the earlier theories, but the basic observation of Henckels, subsequently extended by Julien, that a link between grain size, or texture, and stability exists, has been demonstrated time and again by more recent investigators. The beginning of the present century saw considerable advances in the understanding of the chemistry of metallic sulphides generally. Allen and Crenshaw[41] describe a simple chemical test invented by Stokes to distinguish between pyrite and marcasite, which was later improved and refined by Bannister.[42]

Buehler and Gottschalk[43] attempted to show that pyrite and other mineral sulphides behaved electrochemically with respect to each other, pre-empting current work on both natural 'gossan' weathering and low-temperature electrochemical sulphide oxidation research.

In the latter part of the nineteenth century investigations into the behaviour of pyrite and marcasite in aerated water and air were carried out because pyrite oxidation in mines (especially coalmines) had been strongly suspected as the major cause of fires and explosion.[44] It was shown that powdered pyrite absorbed between 40 and 60 times its own volume of oxygen at $20°C$ in about 100 hours, and produced about 18 joules $m^{-3} \times 10^{-6}$ absorbed oxygen. The rate of oxidation decreased with time, due to the coating of the particles with iron sulphate; removal of the sulphate restored the initial rate of oxidation. Increases in temperature, oxygen pressure and pyrite surface area also increased the oxidation rate. The presence of moisture was necessary for rapid oxidation of crushed pyrite,[45] suggesting that moisture absorption rather than wetting might be a crucial factor in the kinetics of the oxidation.

Because of the application of pyrite oxidation research to the leaching processes of extractive metallurgy and its role in the production of acid streams and mine waters, later research has been largely concerned with the behaviour of pyrite under aqueous conditions. The findings from this type of work are not directly applicable to specimens in collections, but it is essential to review the field in order to gain a consensus view. Nelson *et al.*[46] showed that percolating aerated water oxidized pyrite at a rate inversely proportional to particle size.

6.2.1 Bacterial complicity in pyrite oxidation

In 1919 Powell and Parr[47] stated that the rate-controlling factors for pyrite oxidation in coals were time and bacteria. But it was not until the late 1940s that bacteria were definitely implicated in the production of acid mine waters. Leathen *et al.*[48] isolated a microorganism, *Ferrobacillus ferro-oxidans*, from acid streams, which not only had the ability to oxidize Fe^{2+} to Fe^{3+} but also accelerated the aqueous oxidation of certain forms of crushed pyrite.[49] Silverman and Erlich[50] stated that the particle size and possibly pyrite crystal structure were important factors affecting the rate of oxidation by bacteria. Deschamps and Temple[51] produced evidence that incomplete oxidation of the sulphide moiety resulted in the formation of elemental sulphur, which could then be oxidized to sulphuric acid by *Thiobacillus thio-oxidans*. Kelly and Tuovinen[52] demonstrated that *Ferrobacillus ferro-oxidans* was in fact a thiobacillus, and proposed the name *Thiobacillus ferro-oxidans*. Other types of bacteria, such as *Sulfolobus acidocalderius*, are also known to contribute to pyrite oxidation.

Over the past 40 years there has been controversy about the mechanism of bacterial oxidation of pyrite. While some authorities maintain[53] that thiobacteria simply accelerate the slow but purely chemical oxidation of pyrite, others[54] suggest that thiobacteria can destabilize pyrite direct.

Bacterial complicity in the oxidation of pyrite in museum collections had been considered a possibility during the 1920s. Oemichen[55] was one of the first to propose bactericidal treatment for pyritic museum specimens, suggesting the use of hydroquinone solution. Howie,[3] Pearl[5] and Rixon[56] discussed the possible bacterial complicity in detail. Several proprietary antiseptics were used in attempts to stabilize oxidized material, including 'Dettol', 'Savlon', 'Bacitracin', 'Cetrimide'[57] and 4,chloro-m-cresol.[58] The possible role of bacteria in the oxidation of museum pyrite has been extensively reviewed.[3,10] Unsuccessful attempts to culture thiobacteria from oxidizing museum specimens[59] suggested that neither *T. ferro-oxidans* nor other species of thiobacteria played any part in pyrite oxidation in collections.

6.2.2 Pyrite oxidation under aqueous conditions

Stenhouse and Armstrong[60] showed that crushed pyrite could be oxidized in alkaline solutions at temperatures between $100°C$ and $200°C$ to hydrated iron oxides, with complete loss of the sulphide moiety (which was oxidized to sulphur). Under acid conditions at similar temperature McKay and Halpern[61] demonstrated that oxidation was dependent upon oxygen pressure, and pyrite particle size and temperature, and that the products were iron sulphates and sulphuric acid.

Bailey and Peters[62] studied the breakdown of crushed pyrite in acid solutions between 85° and

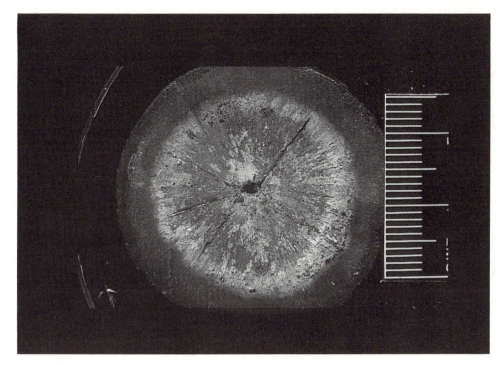

Figure 6.3 Pyrite nodule showing oxidation band, Lower Cretaceous, UK (scale mm).

130°C and between 0 and 0.66 kPa O_2 pressure, and proposed an electrochemical dissolution mechanism where pyrite was anodically corroded by water, which initially formed acid and sulphates; the cathodic reaction reduced the oxygen to water.

Lowson[63] and Goldhaber[64] reviewed mechanisms for aqueous pyrite oxidation at low temperatures. The latter showed that thiosulphate and similar oxyanions were produced during pyrite oxidation at pH 6–9 at 30°C. McKibben and Barnes[65] described the oxidation of pyrite at 10^{-1} kPa in oxygenated acid solution in pH range 2–4, suggesting that oxidation is centred on reactive sites of high excess surface energy, for example, grain edges, corners, defects, cleavages and fractures. Taylor *et al.*[66] presented data on the isotopic composition of sulphates produced during the oxidation of pyrite, which indicated the extent to which both water and oxygen are involved during *in situ* conversion of pyrite to sulphates, either chemically or by bacteria.

The oxidation of sulphide ore bodies has been extensively studied recently. Thornber[67] considers that ore body sulphides, including pyrite, undergo electrochemical oxidation, where both oxygen and galvanic reactions between different sulphides produce gossans, which are the alteration zones commonly found capping sulphide ore bodies. The environmental conditions found in sulphide alter-

ation zones, sulphide-rich tailings, coalmines and other mines in sulphide-rich areas vary considerably from fully saturated to arid. The mechanisms for pyrite oxidation will also vary from bacterial under fully aqueous conditions, through electrochemical in partially saturated conditions, to molecular in dry conditions. In terms of environmental conditions, collections of geological material in museums and other institutions are usually stored under relative humidities which vary between 30% and 80%, i.e. dry to partially saturated conditions.

6.3 Mechanisms for the oxidation of pyrite and marcasite in collections

One of the most striking characteristics of pyrite and marcasite oxidation in collections is its apparent random occurrence. Bannister[2,15] prepared extensive lists of localities which yielded unstable pyrite and marcasite. Howie[3,68] linked the physical structure of pyrite with storage environment, showing that unstable pyrite was usually microcrystalline or framboidal, and that oxidation did not occur substantially until storage relative humidity reached about 60%. Pyrite and marcasite stability may also be influenced by intimate association with carbonaceous material[5,59,69] and association with other sulphides,[70,71] although

Figure 6.4 Chequerboard pattern of tarnish development on polished mount of otherwise stable pyrite, Lower Lias, Dorset. (×175)

none of these exclusively determine the stability of these minerals under storage conditions.

Pyrite and marcasite specimens and material containing one or both of the sulphides can, as we have seen earlier, be loosely classified as inert (or stable) and reactive (or unstable). Inert specimens are often well crystallized aggregates or single crystals. Some well-crystallized specimens will undergo rapid dulling or iridescent tarnishing. Reactive pyrite is rarely massively-crystallized and undergoes more or less rapid conversion to hydrated sulphates and sulphuric acid. Many pyritic fossils, pyrite nodules and ground mass matrices contain pyrite or marcasite which react in this way.

6.3.1 Tarnish formation

The formation of tarnish on well-crystallized pyrite, which ranges from slight dulling of lustre to brown or orange coatings, has been investigated.[72,73] The invisible oxidized mono-layer on pyrite surfaces is probably a pyrite oxysulphur compound rather than an adsorbed oxygen layer. This type of layer, possibly a precursor to the development of a true oxide film, has been considered as a passivating film. Burkin[74] describes the formation of 'passivating' films of oxide

or oxide–hydroxyl complexes on sulphides and metals, under a variety of conditions.

Tarnish develops rapidly on polished specimens of euhedral intergrown stable pyrite (often initially iridescent but turning orange or brown after a few days exposure to air) by differential adsorption of oxygen and/or water vapour on to the differently orientated faces of the sulphide produced by polishing (see Figure 6.4), showing a chequerboard pattern of tarnish, which at high magnification appears as more or less dense spotting, depending upon which polished face it has developed. Pyrite, in common with many other sulphides, exhibits different physical properties and characteristics from face to face, e.g. thermo-EMF, conductivity and resistivity.[75] Metal crystals and synthetic sulphides show variation in surface adsorption properties from face to face, and it would be reasonable to suppose that marcasite and pyrite share this characteristic in terms of oxidizability of different faces.

6.3.2 Pyrite oxidation at high relative humidity levels

Much of the experimental work on pyrite oxidation has been carried out with crushed material, usually under fully aqueous conditions. Crushed samples used in aerial oxidation experiments are acid-washed to remove contamination by accessory minerals and old oxidation products, which could jeopardize the findings. Steger and Desjardins[76,77] demonstrated that, at 52°C and 62% RH, crushed pyrite oxidized via a series of reactions involving transitory formation of ferrous thiosulphate and subsequently ferric sulphate. Lodha *et al.*[78] showed that the storage of pyrite ore in unoxidized conditions could be effectively achieved by keeping RH less than 50% in an atmosphere of nitrogen. Gupta and Singh[79] carried out Mossbaeur studies of pyrite weathering, and concluded that carbonaceous material and particle size were the determining factors in the oxidation of pyrite *in situ*. Banerjee,[80] Carrucio[81] and Howie[3,10] considered that pyrite grain size was the important factor, with framboidal and microcrystalline aggregates of pyrite being more susceptible to oxidation than larger crystallized pyrite masses.

Under fully aqueous conditions the oxidation products of pyrite have two possible fates. Under chemically controlled oxidation conditions, such as occur in acid solutions,[61,71] the oxidation products are immediately removed. The reaction may be effectively first order with respect to oxygen partial pressure, and independent of pH level and ferrous, ferric or sulphate iron concentrations. The rate-controlling process is the reaction of a second molecule of oxygen in the sequence:

$$FeS_2 + O_2 \text{----} FeS_2O_2 \text{ (ads) .... fast}$$

$$FeS_2.(ads) + O_{2\text{----}Fe2}.2O_2 \text{ (ads)----} FeSO_4 + S \text{ .... slow}$$

The rate law for this process is $-d(FeS_2)/dt = kA_{FeS2}P_{02}$, where A_{FeS2} is the surface area of pyrite and k is the apparent rate constant.[82]

Under non-acidic or alkaline conditions oxidation products such as Fe_2O_3 can build up on the pyrite particles,[74,82] effectively slowing pyrite oxidation. The rate of oxidation here is effectively diffusion-controlled.

Pyrite nodules washed out of sediments by the sea appear to survive unchanged for long periods of time, and indeed pyrite sand produced by erosion is abundant in littoral zones. The particles are always, however, rapidly coated by a brown or black layer of oxide. A diffusion-controlled kinetic model suggests that pyrite under these conditions may last for several years.

In pyrite oxidation under the environmental conditions found in collections, water which is in contact with pyrite will play the major role in determining oxidation reaction rates.

Moisture and oxygen absorption isotherms for naturally occuring, untreated sedimentary pyrite

indicate that moisture is a key factor in the oxidation of reactive pyrite (as it probably is for marcasite). Figure 6.5 shows typical moisture adsorption isotherm for natural microcrystalline pyrite together with pyrite oxidation extent after 30 days exposure to various relative humidities. For comparison, Figure 6.6 shows

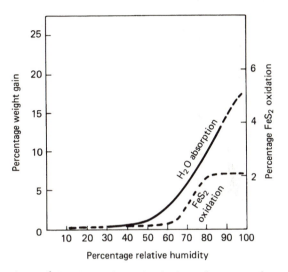

Figure 6.5 Moisture absorption isotherm for untreated nodular framboidal/microcrystalline pyrite after 30 days' exposure to relative humidities of 30–80 per cent with the resultant percentage FeS_2 oxidation.

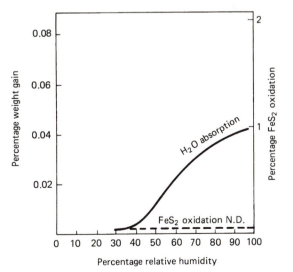

Figure 6.6 Moisture absorption isotherm for untreated stable pyrite after 95 days' exposure to relative humidities of 30–80 per cent.

Figure 6.7 Development of crystalline oxidation products (a–d) on SEM mount of reactive pyrite over approximately 21 days. (Field of view approximately 1000µm)

(a)

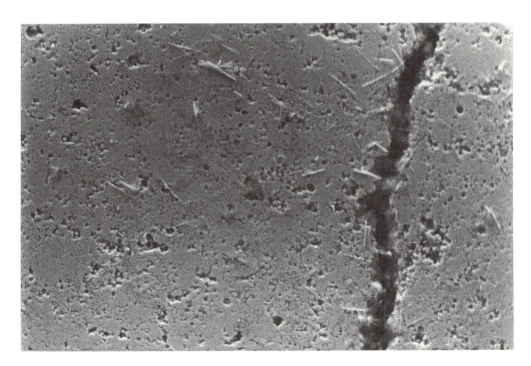

(b)

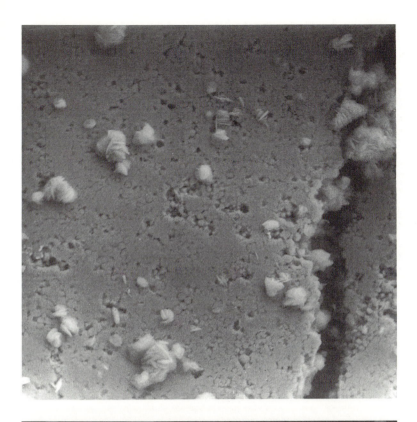

(c)

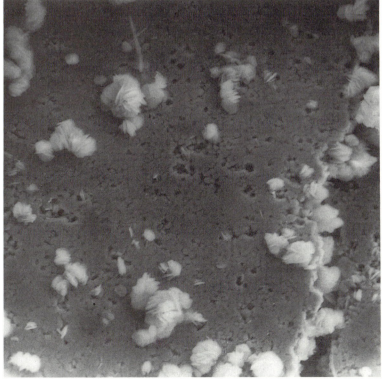

(d)

Figure 6.8 Sites for growth of sulphates on polished reactive pyrite sample (SEM, magnification 2,000).

low reactivity of a stable pyrite. Moisture from air will be adsorbed by most porous materials, the quantity adsorbed being dependent upon the water vapour pressure of the air and the specific surface of the sorbent (that is the total area presented to the sorbing vapour). Wexler[83] and others have shown that inert sorbents will adsorb sufficient moisture to form a mono-layer of water between approximately 0% and 30% to 60%; above a figure between 30–60% and 90% RH water will be adsorbed as a multi-layer.

With reactive sulphides oxygen adsorption will also occur rapidly. Oxygen is physically adsorbed on to the surface of fresh pyrite as soon as it is exposed to the atmosphere. Heating pyrite in air to 700°C causes rapid oxidation, initially to ferrous sulphate and at higher temperatures to sulphur dioxide and ferric oxide. At normal temperatures these oxidation reactions are extremely slow or self-limiting so long as water vapour is not present. Between 30% and 60% RH sufficient water is present to allow partially aqueous oxidation reactions to occur within the multi-layer of water on the surface of pyrite or marcasite.

Morth and Smith[84] found the rate of oxidation increased with increasing RH between 31% and 96% at 24–45°C. When their data was mathematically

treated to give a measure of corrosion sensitivity analogous to that used for metallic oxidation by Waller,[85] the calculated oxidation rate for pyrite doubles for every 26% increase in RH. This suggests that the critical relative humidity for pyrite may well be much lower than 60%. At relative humidities of less than 30% reactive pyrite will, however, probably remain unchanged.[86]

The growth of sulphate product on freshly polished unstable microcrystalline pyrite surfaces undergoing oxidation can be monitored by means of scanning electron microscopy. Figure 6.7 illustrates the progressive formation of uncharacterized hydrated iron sulphates at 75% RH over a period of 21 days. The sites for growth of oxidation product include grain edges and intergranular boundaries; no growth is seen on the faces or cut surfaces of micro-crystals (see Figure 6.8).

After 2 months at 75% RH the sample illustrated was completely oxidized. The rate of growth of sulphate indicated that ions were being transferred via the water multi-layer pervading the porous pyrite aggregate from within the sample.

The mechanisms responsible for the decomposition of pyrite have received considerable renewed attention over the past few years – in particular the electrochemical dissolution of pyrite under aqueous conditions at low temperatures (295°K) and normal atmospheric pressure.[87,88] In addition, the semiconductor properties of pyrite are now considered to be of significance with respect to elecro-oxidation susceptibility.[89] The extent to which close contact with other sulphides inhibits or accelerates pyrite oxidation,[90] and diffusion of sulphur into other sulphide phases, adds new dimensions to the chemistry of pyrite decomposition.

The electrochemical contribution to the oxidation of pyrite under typical storage conditions has received very little attention. Preliminary research,[91] however, indicates that electrode dissolution reactions of the type:

$$FeS_2 + 8H_2O \rightarrow FeSO_4 + H_2SO_4 + 14H^+ + 14e^-$$
(anodic half reaction)

$$O_2 + 4H^+ + 4e \rightarrow H_2O \text{ (cathodic half reaction)}$$

occur, perhaps accounting for 50% of the oxidation mechanism. Very little work has been done on the oxidation of marcasite recently, but grain size and surface area for reaction dictate susceptibility to oxidation in the same way as for pyrite.

6.3.3 Pyrite and marcasite oxidation products

The basic oxidation reaction of pyrite or marcasite to yield sulphuric acid and ferrous sulphate is somewhat misleading in reality. Hydrated ferrous sulphate will itself oxidize, hydrolyse or dehydrate, depending upon environmental conditions, to yield products ranging from ferrous sulphate (low to high hydrates), ferroso-ferric sulphate-hydrates, basic sulphate-hydrates, ferric sulphates (low to high hydrates) to hydrate phases containing free sulphuric acid.[92,93,94] The acid-sulphate mixtures commonly attack associated minerals, producing a variety of sulphates, often mixed or multi-cation phases.

Merwin and Posnjac[95] observed that the complete oxidation of pyrite in mine situations resulted in acid ferric sulphates and rhomboclase. The sulphates closest to the pyrite were the most acidic and least oxidized, while at the surface of the crystalline mass of oxidation products the sulphates were more basic and highly oxidized.

In collections, oxidized pyrite and marcasite usually yield, first, melanterite ($Fe_2O_4.7H_2O$), which, on further oxidation, produces copiapite (Fe^{II}, Fe^{III}_4 ($SO_4)_6(OH)_2$), fibroferrite and other hydrates. The range of oxidation products is extensive and very much dependent on minerals or matrixes associated with oxidizing pyrite or marcasite.[92,96]

6.4 Conserving pyrite and marcasite

A historical review of methods used for treating pyrite and marcasite[3] revealed that the earlier methods relied on isolating specimens from the atmosphere, using either fluids such as linseed oil, petroleum or paraffin or consolidants such as shellac or wax. Bather[97] proposed that specimens should be neutralized in hot caustic alkali solution, and alcohol-dried before consolidation. Radley[14] recommended washing and air-drying of specimens, followed by cellulose nitrate consolidation. Later methods utilized ammonium hydroxide vapour for neutralization, and various consolidants, including polyvinyl acetate,[15] Butvar,[68] epoxy resins,[98] and polybutyl methacrylate.[56] Other types of treatment have included the use of concentrated hydrochloric acid[99] to remove oxidation products, vapour phase corrosion inhibitors and, most recently, combined neutralization/oxidation product-removal treatments such as aqueous sodium bicarbonate,[100] ammonium thioglycollate in alcohol[10] and ethanolamine thioglycollate.[101]

The key to successful treatment of freshly collected, unoxidized pyrite and marcasite specimens, and minerals associated with either of these sulphides, is that they should not be treated with any water-based cleaning agents, acids or alkalis. Dry-cleaning or cleaning using organic solvents followed by storage at 30% RH is recommended. Few localities yield completely stable specimens, and it is safest to assume that all pyritic material and marcasite are susceptible to oxidation, and to treat them accordingly. Removal of oxide films or other minerals which form part of the paragenesis (which may be a secondary paragenesis consisting of weathering or alteration products) should be avoided unless the material is required for analysis. Here the chemical treatment chosen must make allowance for any later analytical work, i.e. reduce potential contamination to zero.

Full documentation of specimen conditions is an essential part of the conservation process. Data recorded should include details of collection locality, processing methods (if any), storage conditions (climate records if available), condition photographs, analytical data (e.g. XRD determinations, etc.) on any oxidation products, treatment schedules, future action and condition review.

In order to ensure that oxidation does not begin in the more reactive material, the fundamental requirement is storage at low relative humidity – as much below 60% RH as possible. Oxidation reactions may occur with pyrite or marcasite, in certain associations even at low RH, e.g. lignitic pyrite, and fine grained pyrite associated with minerals such as galena or sphalerite. Packing reactive types of pyrite in polythene or Saran bags for long-term storage will usually trap sufficient moisture to set off oxidation reactions within 1 to 2 days, and is not recommended unless the specimens have been vacuum-dried and vacuum-packed in the bags.

The recommended course of action with this type of material is rapid transfer to the laboratory to a desiccator, or, if this is not available, transfer to an organic solvent such as dry acetone or dry isopropanol, for one or two days to remove absorbed water and thence to dry over silica gel. Pyrite or marcasite should not be oven-dried, because as soon as the warm mineral is removed to cooler surroundings, it will rapidly absorb moisture from the air as it cools.

Should there be any ferrous or ferric sulphates on the pyrite or marcasite as a result of oxidation before retrieval from the field or from oxidation in transit, these products may, if starved of water vapour, soon be reduced. In closed systems sulphuric acid and ferric sulphates will react with pyrite to produce more and more basic salts until iron oxides are produced. These reactions will probably be self-limiting within a few days, and should not cause any visible change. Hydrated ferrous sulphates, however, may well remain as a surface bloom, which may later respond to changing relative humidity and help to reactivate the oxidation process.[3]

All pyritic material should be stored in areas where relative humidity can be controlled to 50% or less. See Thomson[102] for methods of achieving desired relative humidities. Small cases and individual boxes if well sealed can be kept at humidities of 30% to 40% for long periods, using dry silica gel. Larger cases and cabinets need to be stored in controlled environment conditions, such that external RH stays below 60% at all times throughout the year.

The conservation treatment of oxidizing specimens will depend, first, upon the extent of deterioration and, second, on the importance of the specimen. Extensively oxidized material is rarely worth conserving. Howie[3,68] reviewed many of the methods used and concluded that few were very effective. The use of polymers should be considered only for friable specimens as support. Vacuum impregnation of sulphides, using polymers such as polyvinyl acetate, acrylics, etc., will only be effective if it can be ensured that the consolidant penetrates the specimen. None of the organic polymers routinely used in conservation can as yet be guaranteed impermeable to oxygen and, more importantly, water vapour[103].

References

1 HEY, M.H., *An Index of Mineral Species and Varieties*, with two appendices, 1963 and 1974, British Museum, Natural History, London (1975)

2 BANNISTER, F.A., 'The preservation of species of marcasite and pyrite', *Museums Journal*, 33, pp. 72-5 (1933)

3 HOWIE, F.M.P., 'Pyrite and conservation', *Newsletter of the Geological Curators Group*, 1, pp. 457-65 (1977)

4 WALLER, R., 'The preservation of mineral specimens', *Preprints of the AIC Meeting*, San Francisco, pp. 116-28 (1980)

5 PEARL, R.M., *Cleaning and Preserving Minerals* Earth Science Publishing Co., Colorado Springs (1975)

6 STARLING, K., 'Deterioration of medieval coal at high relative humidity', *Conservation News*, 22, p. 13 (1983)

7 SHAYEN, A., 'Deterioration of a concrete surface due to the oxidation of pyrite contained in pyritic aggregates', *Cement and Concrete Research*, 18, pp. 723-30 (1988)

8 JANGDAHL, C.E., 'Swelling shale in the Ostersund area', *National Swedish Building Research Report*, R35 (1971)

9 STEGER, H.F., 'The stability of the certified reference ores MP-1, KC-1 and SU-1 toward air oxidation', *Talanta*, 23, pp. 643-8 (1976)

10 HOWIE, F.M.P., 'Storage environment and the conservation of geological material', *Conservator*, 2, pp. 13-19 (1978)

11 LODHA, T.R., SINHA, N.A. and SRIVASTAVA, A.C., 'Storage problems of Amjhore pyrites', *Fertiliser Technology*, 20, pp. 76-7 (1983)

12 CARRUCIO, F., 'Trace element distribution in reactive and inert pyrite', *Proceedings 4th Symposium on Coal Mine Drainage, Research Preprints*, pp. 48-53 (1972)

13 BUCKLEY, A.N., HAMILTON, I.L. and WOODS, K., 'Studies of the surface oxidation of pyrite and pyrrhotite using x-ray photoelectron spectroscopy and linear sweep voltammetry', *Proceedings Electrochemical Society*, pp. 88-121 (1988)

14 RADLEY, E.G., 'The decomposition of pyritized and other fossils', *The Naturalist*, pp. 143-6, 167-73, 196-202 (1929)

15 BANNISTER, F.A., 'The preservation of minerals and meteorites', *Museums Journal*, 36, pp. 465-76 (1937)

16 DANA, E.S., *The system of mineralogy of James Dwight Dana*, Wiley, New York and London (1892)

17 WALLERIUS, J.G., *Mineralogie*, Paris (1753)

18 HILL, J., *Fossils Arranged*, London (1771)

19 CRONSTEDT, A., *An Essay towards a System of Mineralogy*, Translated by G. van Engenstrom, London (1788)

20 DAVIES, A.M., *An Introduction to Palaeontology*, Allen and Unwin, London (1961)

21 SHEPHERD, W., *Flint, its Origins, Properties and Uses*, Faber and Faber, London (1972)

22 SINKANKAS, J., *Gemstone and Mineral Data Book*, Winchester Press, New York (1972)

23 EICHHOLZ, D.E., *De lapidibus*, Translation, Oxford University Press, Oxford, (1965). Original by Theophrastus, 315-14 BC

24 PLINIUS SECUNDUS, C., *Natural History*, Translated by H. Rackham, *et al.*, 10 Volumes, William Heinemann, London (1958-62)

25 AGRICOLA, G., *De Re Metallica*, Translated by Hoover, H.S. London (1912)

26 MAYOW, J., see Partington, R.A., *History of Chemistry*, Vol. II, Macmillan, London (1969)

27 BOYLE, R., see Partington, R.A., *History of Chemistry*, Vol. II, Macmillan, London (1969)

28 LAVOISIER, A., Mémoires sur la vitriolization de pyrites martiales, *Memoirs Academie des Sciences*, Paris (1789)

29 HENCKELS, J.F., *Pyritologia*, English translation, London (1754)

30 WERNER, J., see Partington, R.A., *History of Chemistry*, Vol. II, Macmillan, London (1969)

31 BERZELIUS, J.J., see Partington, R.A., *History of Chemistry*, Vol. II, Macmillan, London (1969)

32 HAUSMANN author's data

33 HATCHETT, C., 'An Analysis of the magnetical pyrites: with remarks on some of the other sulphurets of iron', *Philosophical Transactions of the Royal Society*, 94, pp. 315-45 (1804)

34 PROUST, L.G., 'Sur quelques sulfures métalliques', *Journal de Physique de Chemie et d'Histoire Naturelle*, 53, pp. 89-97 (1801)

35 KOHLER, F., Ueber den Strahlkies von Grossallmerode in Hessen, Annalen Chimie. Physiks, 14, p. 91-6 (1828)

36 NICOL, J., *Manual of Mineralogy*, Edinburgh (1849)

37 SENFT, K.F.E., *Die Krystallinishen Felsgamengtheile*, Berlin (1868)

38 KIMBALL, J.P., Relations of sulphur in coal and coke: atmospheric oxidation or weathering of coal', *Transactions American Institution of Mining Engineers*, 8, pp. 181-225 (1879-80)

39 JULIEN, A.A., 'On the variation of decomposition in the iron pyrites; its cause and its relation to density, *Annals of the New York Academy of Science*, 3, pp. 365-404 (1885)

40 JULIEN, A.A., 'On the variation of decomposition in the iron pyrites; its cause and its relation to density', *Annals of the New York Academy of Science*, 4, pp. 133-228 (1889)

41 ALLEN, E.T. and CRENSHAW, J.L., 'The Stokes method for the determination of pyrite and marcasite', *American Journal of Science*, 38, pp. 371-92 (1914)

42 BANNISTER, F.A., 'The distinction of pyrite from marcasite in nodular growths', *Mineral Magazine*, 33, pp. 179-87 (1932)

43 BUEHLER, H.A. and GOTTSCHALK, V.M., 'Oxidation of sulphides', *Economic Geology*, 7, pp. 371-92 (1912)

44 WINMILL, T.F., 'The atmospheric oxidation of iron pyrites and the absorption of oxygen by coal', *Transaction of the Institute of Mining Engineers*, 51, pp. 500-47 (1916)

45 LI, S.H. and PARR, S.W., 'The oxidation of pyrites as a factor in the spontaneous combustion of coal', *Industrial and Engineering Chemistry*, 25, pp. 1299-1304 (1933)

46 NELSON, H.W., SNOW, R.D. and KEYES, D.B., 'Oxidation of pyritic sulphur in bituminous coal', *Industrial and Engineering Chemistry*, 25, pp. 1355-8 (1933)

47 PARR, S.W., *Effects of storage upon the Properties of Coal*, Illinois University of Engineering and Experimental Station, Bulletin, 19 (1917)

48 LEATHEN, W.W., BRALEY, S.A. and MCINTYRE, L.D., 'The role of bacteria in the formation of acid from certain sulphuric constituents associated with bitumous coal. II. Ferrous iron oxidizing bacteria', *Applied Microbiology*, 2, pp. 65-8 (1953)

49 BYNER, L.C. and ANDERSON, R., 'Microorganisms in leaching sulphide minerals', *Industrial Engineering Chemistry*, 49, pp. 1721-4 (1958)

50 SILVERMAN, M.P. and ERLICH, H.E., 'Microbial formation and degradation of minerals', *Advances in Applied Microbiology*, 6, pp. 153-206 (1964)

51 DESCHAMPS, E.H. and TEMPLE, K.L., 'Autotropic bacteria and the formation of acid in bituminous coal mines', *Applied Mineralogy*, 1, pp. 255-8 (1953)

52 KELLY, D.P. and TUOVINEN, O.H., 'Recommendation that the names *Ferrobacillus thiooxidans* Leathen and Braley and *Ferrobacillus sulfoxidans* Kinsel be recognized as the synonyms of *Thiobacillus ferrooxidans* Temple and Colmer', *International Journal of Systematic Bacteriology*, 22, pp. 170-2 (1972)

53 BRYNER, L.C., BECK, J.F., DAVIS, D.B. and WILSON, D.G., 'Microorganisms in leaching sulphide minerals', *Industrial and Engineering Chemistry*, 46, pp. 2578-92 (1954)

54 BECK, J.F. and BROWN, E.G., 'Direct sulphide oxidation in the solubilization of sulphide ores by *Thiobacillus ferrooxidans*', *Journal of Bacteriology*, 96, pp. 1433-4 (1968)

55 OEMICHEN, E., 'Traitement d'anti-oxydation des fossiles puyriteux', *Comptes Rendue Sommaire Seanc Société Géologique Français*, 6, pp. 65-7 (1944)

56 RIXON, A.E., *Fossil Animal Remains: Their Preparation and Conservation*, Athlone Press, London (1976)

57 BROADHURST, F.M. and DUFFY, L., 'A plesiosaur in the Geology Department University of Manchester', *Museums Journal*, 70, pp. 3-31 (1970)

58 BOOTH, G.H. and SEFTON, G.V., 'Vapour phase inhibition of thiobacilli and ferrobacilli: a potential preservation for museum specimens', *Nature*, 226, pp. 30-1 (1970)

59 DACEY, P., personal communication to author (1976)

60 STENHOUSE, J.F. and ARMSTRONG, N.M., 'The aqueous oxidation of pyrite', *Canada Mining and Metallurigcal Bulletin*, 45, pp. 48-53 (1952)

61 MCKAY, D.R. and HALPERN, J., 'A kinetic study of the oxidation of pyrite in aqueous suspension', *Transactions of the Metallurgical Society of AIME*, 212, pp. 301-9 (1958)

62 BAILEY, L.K. and PETERS, E., 'Decomposition of pyrite in acids by pressure leaching and anodization: the case for an electrochemical mechanism', *Canadian Metallurgical Quarterly*, 15, pp. 333-4 (1976)

63 LOWSON, R.T., 'Aqueous oxidation of pyrite by molecular oxygen', *Chemistry Review*, 82, pp. 401-97 (1982)

64 GOLDHABER, M.B., 'Experimental study of metastable sulphur oxyanion formation during pyrite oxidation at pH 6-9 and 30°C', *American Journal of Science*, 283, pp. 193-217 (1983)

65 MCKIBBEN, M.H. and BARNES, H.L., 'Oxidation of pyrite in low temperature acidic solutions: rate laws and surface textures', *Geochemica et Cosmochemica Acta*, 50, pp. 1509-20 (1986)

66 TAYLOR, B.E., WHEELER, M.C. and NORDSTROM, D.K., 'Isotopic composition of sulphate in acid mine drainage as a measure of bacterial oxidation', *Nature*, 308, pp. 538-41 (1984)

67 THORNBER, M.R., 'Supergene alteration of sulphides VII: distribution of elements during the gossan forming process', *Chemical Geology*, 53, pp. 279-301 (1985)

68 HOWIE, F.M.P., 'Storage and exhibition environment and the conservation of fossil material', *Preprint Museums Conservation Climate Conference*, ICCROM, Rome (1978)

69 SCHOPF, J.M., EHLERS, E.G., STILES, D.V. and BIRKE, J.D., 'Fossil iron bacteria preserved in pyrite', *Proceedings of the American Philosophical Society*, 109, pp. 288-308 (1965)

70 PALACHE, C., BERMAN, H. and FRONDEL, C., *Dana's System of Mineralogy Elements, sulphides, sulphosalts and oxides*, Wiley, New York (1944)

71 PETERS, E. and MAJIMA, H., 'Electrochemical reactions of pyrite in acid perchlorate solutions', *Canadian Metallurgy Quarterly*, 7, pp. 111-17 (1968)

72 MITCHELL, D. and WOODS, R., 'Analysis of oxidized layers on pyrite surfaces by x-ray emission spectroscopy and cyclic voltammetry', *Australian Journal of Chemistry*, 31, pp. 27-34 (1978)

73 BUCKLEY, A.N. and WOODS, R., Photoelectron spectroscopic study of the oxidation of chalcopyrite, *Australian Journal of Chemistry*, 37, pp. 2403-13 (1984)

74 BURKIN, A.R., *The Chemistry of Hydrometallurgical Processes*, The Spon Press, London, (1966); and 'Pyrite and oxidation in alkali: *Solid-state* transformations during leaching', *Mineral Science Engineering*, 1, pp. 4-14 (1969)

75 MELLOR, J.W., *Treatise on Inorganic Chemistry*, Longman, London (1933)

76 STEGER, H.F. and DESJARDINS, L.E., 'Oxidation of sulphide minerals III: determination of sulphate and thiosulphate in oxidized sulphide minerals', *Talanta*, 24, pp. 675-9 (1977)

77 STEGER, H.F. and DESJARDINS, L.E., 'Oxidation of sulphide minerals IV: pyrite, chalcopyrite and pyrrhotite', *Chemical Geology*, 23, pp. 225-37 (1978)

78 LODHA, T.R., SINTIA, N.K. and SRIVASTAVA, A.C., 'Storage problems of Amjhore pyrites', *Fertilizer Technology*, 20, pp. 76-7 (1983)

79 GUPTA, V.P. and SINGH, O.P., 'Weathering effects on pyrite FeS_2', *Indian Journal of Pure and Applied Physics*, 22, p. 626 (1984)

80 BANERJEE, A.C., 'Mechanisms of oxidation of iron pyrites', *Chemical Society of London, Chemical Communications Journal*, pp. 1006-7 (1971)

81 CARRUCIO, F.T., 'The quantification of reactive pyrite by grain size', *Proceedings 3rd Symposium on Coal Mine Drainage Research, Preprints*, pp. 123-31 (1970)

82 RICKARD, D., 'Reaction kinetics in low temperature sulphide ore formation: principles and applications', (MS paper)

83 WEXLER, A., *Moisture and Humidity*, Vol. 3, Reinhold, New York (1965)

84 MORTH, A.H. and SMITH, E.E., 'Kinetics of the sulphide to sulphate reaction', *American Chemical Society; Division of Fuel Chemistry*, 10, pp. 83-92 (1966)

85 WALLER, R., 'An experimental ammonia gas treatment method for oxidized pyrite mineral specimens', *Triennial Report. ICOM Committee for Conservation*, pp. 625-30 (1987)

86 WALLER, R., 'Pyrite Oxidation Studies', *Newsletter Canadian Conservation Institute Spring/Summer*, p. 10 (1989)

87 LALVANI, S.B. and SHANI, M., 'Indirect electro-oxidation of pyrite slurries: reaction rate studies', *Chemical Engineering Communications*, 70, pp. 215-25 (1988)

88 ORLOVA, T.A. and STUPNIKOV, V.M., 'Mechanisms of oxidative dissolution of sulphides', *Zschreift Prik Khim*, 61, pp. 2172-7 (1988)

89 MISHRA, K.K., *Dissolution and semiconductor electrochemistry of pyrite*, Pennsylvania State University (1988)

90 ELISEEV, N.I., NEMESHAEVA, L.A. and BORSKOV, F., 'Electrochemical leaching of pyrite', *Electron Obrab Mater*, 6, pp. 38-41 (1988)

91 HOWIE, F.M.P., 'An electrochemical mechanism for the oxidation of pyrite in museum-stored specimens', *Proceedings of the 1985 Workshop on Care and Maintenance of Natural History Collections*, Royal Ontario Museum, p. 28 (1986)

92 BUURMAN, P., '"Invitro" weathering products of pyrite', *Geologie en Mijnbouw*, 54, pp. 101-5 (1975)

93 WALLER, R., this volume (Chapter 3)

94 YOUNG, B. and NANCARROW, P.H.A., 'Rozenite and other sulphate-rich minerals from the Cumbrian coalfields', *Mineralogical Magazine*, 52, pp. 551-2 (1988)

95 MERWIN, H.E. and POSNJAC, E., 'Sulphate encrustations in the Copper Queen Mine Bisbee Arizona', *American Mineralogist*, 22, p. 567 (1937)

96 SCLAR, C.B., 'Decomposition of pyritized shale to halotrichite and melanterite', *The American Mineralogist*, 46, pp. 754-6 (1961)

97 BATHER, F.A., 'The preparation and preservation of fossils', *Museums Journal*, 8, pp. 76-90 (1908)

98 ODDY, W.A., 'The conservation of pyrite stone', *Studies in Conservation*, 22, pp. 68-72 (1977)

99 GORDON, S.G., 'Preservation of specimens of marcasite and pyrite', *American Mineralogist*, 32, p. 589 (1947)

100 HANSEN, G.R., 'Preserving pyrite', *Mineralogical Record*, 10, p. 56 (1979)

101 CORNISH, L. and DOYLE, A., 'Use of ethanolamine thioglycollate in the conservation of pyritized fossils', *Palaeontology*, 27, pp. 421-4 (1984)

102 THOMSON, G., *Museum Environment*, Butterworths, London (1987)

103 BUILDING RESEARCH ESTABLISHMENT, 'Deterioration of Mundic Concrete Houses', *BRE News* (1991)

7

Meteorites

Dr A.W.R. Bevan

Meteorites are fragments of natural extraterrestrial material which survive their fall to Earth from space. The available chemical and astronomical evidence suggests that meteorites originate within the Solar System, and most are pieces of rock and/or metal detached from asteroids in elliptical orbits around the sun. Radiometric dating of meteorites shows that many are 4,550 million years old, and have remained essentially unaltered since their formation. As samples from minor, inactive planets, meteorites are a unique source of information on a wide variety of events associated with the formation and early history of the Solar System. For this reason meteorites are of immense scientific value, and for almost two centuries their study has been associated with both technical and philosophical advances in the disciplines of physics, chemistry, mineralogy and astronomy.

Only five or six meteorites are witnessed to fall world-wide and recovered each year. Since two-thirds of the Earth's surface is covered by water and less than one third of the land surface is populated, most meteorite falls are unobserved, and many land in the oceans. Many meteorites are composed of aggregates of minerals, some of which are unknown on earth and include a few that corrode rapidly in the moist, oxygen-rich terrestrial atmosphere. In temperate and tropical climates, unrecovered meteorites disintegrate on a timescale that is short compared to their rate of infall. However, in 'dry' desert areas of the world meteorites may be preserved, to be found thousands of years after their fall.

About 20 years ago there were, in total, about 2,100 meteorites known to science. Approximately 60% of the meteorites then stored in collections throughout the world were so-called meteorite 'finds'. In 1969 a party of Japanese geologists found nine fragments of meteorites lying on 'blue' ice in

the Yamato Mountains area of Antarctica. The fragments proved to represent four different chemical groups of meteorites, and so belonged to at least four different falls. 'Blue' ice is being actively uplifted and eroded by wind action. Subsequently it was realized that meteorites which had fallen at different times in the past had become incorporated into the ice and been brought to the surface again by the processes of transport and erosion. Although meteorites essentially fall randomly over the surface of the Earth, in Antarctica the unique preservation qualities of the ice provide a means of concentrating meteorites in space and time.

Further expeditions to Antarctica by Japanese and American scientists in 1973 and every year since have discovered numerous sites with conditions similar to those of the Yamato Mountains and an abundance of meteorites. To date, more than 9,800 specimens of meteorites have been recovered from Antarctica.[1] Since many fragments of similar types of meteorite have been found together in small areas, it has yet to be determined how many distinct meteorite falls are represented among Antarctic finds, though it has been suggested that the number is approximately one-tenth of the number of specimens collected.

In recent years, as interest in meteorites has heightened, other desert areas of the world have proved abundant sources of meteorite finds. Large numbers are currently being recovered from the 'bad lands' of New Mexico, the Nullarbor Plain in Australia and, more recently, the stony deserts of North Africa.

7.1 Arrivals from space

During a single year taken to orbit the Sun the Earth sweeps up hundreds of thousands of tonnes of

Figure 7.1 Fragment of the stony (chondritic) meteorite that fell at Barwell, Leicestershire in 1965. The dark fusion crust is composed of oxides of iron.

natural debris from space. The bulk of the material arrives either as small fragments that are destroyed in the upper atmosphere as *meteors*, or as microscopic dust particles that drift down gently on to the Earth's surface.

Even if an object has no motion relative to the Earth, gravitational attraction will cause it to enter the atmosphere at a minimum velocity of 11.2 km/sec (the escape velocity; about forty times the speed of sound). At such hypersonic velocities, frictional heating in the atmosphere causes the surface of the object to melt and vaporize, and the surrounding air to ionize. The resulting body of incandescent gas and dust, called a 'fireball', can give rise to brief, but spectacular visual displays.

Melted material is stripped away rapidly from the surface of the falling body to produce a trail of solidified droplets. This stripping process, known as

ablation, removes molten material so efficiently that heat is not conducted into the interior of the object, which remains cold. Objects large enough to survive ablative flights are slowed down in the dense, lower atmosphere to a point (usually 10–15 km from the surface) where frictional melting and luminous phenomena cease. The remaining melt on the surface of the object solidifies to form a thin 'fusion' crust (Figure 7.1) and the body free-falls to Earth under gravity to become a meteorite.

In the last 25 years several countries throughout the world have established automated networks of cameras to scan selected parts of the sky for the purpose of photographing bright fireballs. Of the hundreds of fireballs that have been recorded in this way, only three have led to the recovery of meteorites (Table 7.1). From the photographic record it has been possible to determine accurate

Table 7.1 Meteorites recovered as the result of fireball photography

Meteorite	Date	Class	Camera network	Measured velocity km/sec	Recovered material
Pribram, Czechoslovakia	April, 1959	ordinary chondrite	Czechoslovakia European	20.9	Four stones, total weight 5.555 kg
Lost City, Oklahoma, USA	January 1970	ordinary chondrite	Prairie Photographic Network	14.2	Four fragments, totalling 17 kg
Innisfree, Alberta, Canada	February, 1977	ordinary chondrite	Canadian Photographic Network	14.54	Fragments totalling 4.58 kg

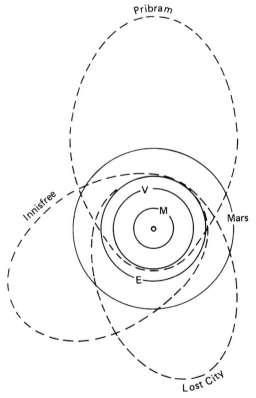

Figure 7.2 Orbits of the bodies that yielded the 'Pribram', 'Lost City' and 'Innisfree' meteorites in relation to the orbits of Mercury (M), Venus (V), the Earth (E) and Mars. All three meteorite orbits are elliptical and extend out beyond Mars, but none intersect the orbit of Venus.

velocities and orbits for the bodies that give rise to meteorites. In each of the three recorded cases the calculated orbits are elliptical and extend out into the broad region between Mars and Jupiter occupied by small asteroids (Figure 7.2).

Recent discoveries have shown that meteoritic materials are not exclusively samples from asteroids. Eleven meteorites recovered from the Antarctic ice have been identified as fragments of the Moon. Additionally, it has been suggested that eight meteorites may have been derived from the planet Mars. However, the majority of meteorites that are observed to fall are undoubtedly of asteroidal origin.

Individual meteoritic masses that have been recovered vary in weight from fractions of a gram up to tens of tonnes. Large, fissile bodies may be disrupted during their flight through the atmosphere to give rise to 'showers' of meteorites that may comprise hundreds or thousands of individuals. Single, undisrupted bodies weighing hundreds of tonnes or more are not slowed appreciably by the atmosphere and impact at hypersonic velocities to produce explosion craters.

During the formation of an explosion crater the bulk of the projectile is destroyed by vaporization and thereby sets an upper limit (approximately 100 tonnes) to the size of meteoritic masses expected to survive as single bodies. Thus, for dynamic reasons, meteorites are not usually found inside explosion craters. In the case of 'Meteor Crater', Arizona, more than 30 tonnes of meteorite fragments, representing residual material from the projectile, have been recovered from the crater rim and surrounding plains.

Cratering events on the scale of 'Meteor Crater' are predicted to occur on Earth once every 25,000 to 50,000 years, and even larger, catastrophic impacts once every 15 million years.

7.2 The nature and origin of meteoritic materials

No attempt is made here to include other than the more general aspects of meteorite taxonomy. There are available today a number of excellent texts on meteorites, to which the reader is referred for a more detailed account (see Further reading). The

Table 7.2 Minerals that commonly occur in meteorites

Silicates	olivine, $(MgFe)_2SiO_4$
	pyroxene, $(MgFeCa)_2Si_2O_6$
	plagioclase feldspar, $NaAlSi_3O_8–CaAl_2SiO_8$
Metallic iron–nickel	kamacite (body-centred cubic metal with less than 7.5 wt% Ni)
	taenite (face-centred cubic metal containing more than 25 wt% Ni)
Sulphides	troilite (mono-sulphide of iron) FeS
	daubreelite, $FeCr_2S_4$
	pentlandite, $FeNi_9 S_8$
Others	serpentine, $Mg_3[Si_2O_5](OH)_4$
	apatite, $Ca_5(PO_4)_3(F,Cl)$
	magnetite, Fe_3O_4
	chromite, $Fe^{+2}Cr_2O_4$
	epsomite, $MgSO_4.7H_2O$
	graphite (carbon)

following is merely a brief review of the essential classification of meteorites, and a description of the nature and origin of meteoritic materials in so far as it bears on special problems associated with their curation and conservation.

Meteoritic materials are dominated by two main types of minerals: silicates rich in iron and magnesium, such as olivine $(FeMg)_2SiO_4$ and pyroxene $(FeMgCa)_2Si_2O_6$, and native metallic iron containing variable amounts of nickel (Table 7.2). Historically, meteorites have been divided into three broad categories on the basis of the relative proportions of ferro-magnesian silicates and metallic iron-nickel that they contain. *Irons* are composed principally of metal; *stones* consist predominantly of silicates but also contain varying subsidiary amounts of metal; and *stony-irons* comprise metal and silicates in roughly equal proportions.

Detailed chemical and mineralogical studies of meteorites have identified numerous sub-divisions within these three categories. The principal classes of meteorite are listed in Table 7.3. Of the two types of stony meteorite recognized, the *chondrites* are most numerous, accounting for 86% of all meteorites observed to fall. Chondritic meteorites have chemical compositions close to that of the non-volatile portion of the Sun, and are characterized by the presence of near spherical, mm-sized bodies of predominantly silicate material called *chondrules* (Greek 'chondros' = grain), from which their name derives.

A rarer group of stony meteorites, the *achondrites*, differs from the chondrites in lacking chondrules and being generally devoid of metallic iron-nickel (Figure 7.3). In general, achondrites resemble terrestrial igneous rocks or their debris.

In detail, chondritic meteorites comprise intricate and complex aggregates of silicate, metallic and sulphur-rich minerals, with accessory oxides and phosphates. With the exception of volatile elements such as hydrogen, carbon, nitrogen and sulphur, all chondritic meteorites have remarkably similar chemical compositions, the elements iron, magnesium, silicon and oxygen constituting over 90% by weight. However, systematic variations in the total content of iron between chondrites and the distribution of this element between reduced (metal and sulphide) and oxidized (silicates and oxides) minerals serve to distinguish a number of groups of chondrites.

The highly reduced, iron-rich *enstatite chondrites* contain abundant metal and sulphide and, as their name suggests, virtually iron-free silicates. In contrast, *carbonaceous chondrites* are highly oxidized materials containing little or no metal and

Table 7.3 Principal meteorite types and their abundance

		Number	Observed falls %	Fall frequency%
Stones				
Chondrites	enstatite	24	54.2	1.6
	ordinary	1,446	45.7	79.7
	carbonaceous	67	52.2	4.2
Achondrites		132	52.3	8.3
Stony-irons				
Pallasites		39	7.7	0.3
Mesosiderites		32	18.8	0.7
Irons		725	5.8	5.0
Total well-documented and classified meteorites		2,465	33.63	

Excluded are 144 authenticated but poorly documented, unclassified or otherwise anomalous stones and two stony-irons. A further 173 'doubtful' meteorites of all types are also excluded. These are either fragments of other documented meteorites, or material whose fall or find was reported but whose authenticity remains unsubstantiated. A limited number of Antarctic meteorite finds (all types) are included (up to and including those collected in 1980–1). Only those chondrites from Antarctica of mass greater than 200 grams are included.

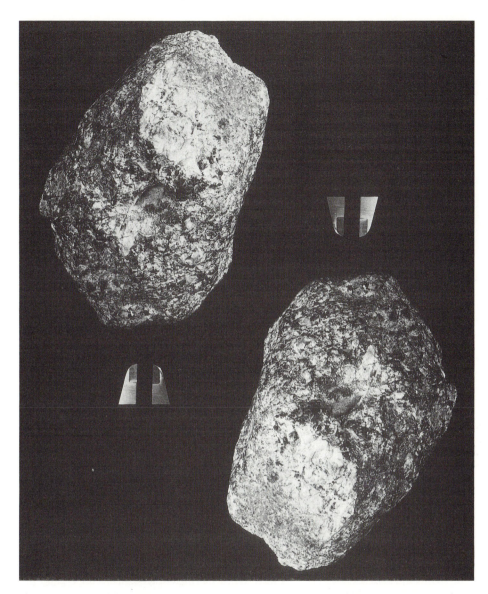

Figure 7.3 Mass of the 'Mayo Belwa' achondritic stony meteorite that fell in Nigeria in 1974. The meteorite, which is composed predominantly of large crystals of enstatite, is one of only eleven in its class (aubrites) known to science (calipers 1 cm apart).

many minerals that formed at low temperatures. Often rich in carbon, both as the free element and as complex carbon compounds, carbonaceous chondrites also contain appreciable amounts of water (up to 20 wt%), silicates such as serpentine that are hydroxyl-bearing, and hydrous sulphates such as epsomite and gypsum.

The *ordinary chondrites* contain significant amounts of both oxidized and metallic iron and are the most abundant extraterrestrial materials available for study.

Iron meteorites contain total amounts of nickel that vary from 5–60 wt%. Most irons contain 7–12 wt% Ni and comprise two cubic metallic minerals (kamacite and taenite) in an octahedral arrangement of interlocking plates called a Widmanstätten pattern (Figure 7.4). Widmanstätten structures are formed by the exsolution of plates of kamacite (body-centred cubic Fe-Ni) from hot, solid crystals of taenite (face-centred cubic Fe-Ni) during cooling on a timescale of millions of years. Iron meteorites that display Widmanstätten patterns are called *octahedrites*. A

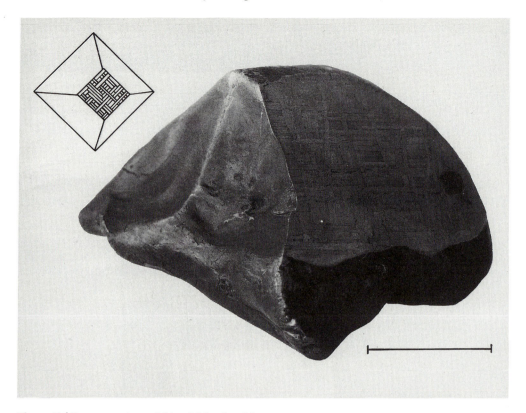

Figure 7.4 Iron meteorite weighing 3.5 kg that fell at Rowton, near Wellington, Shropshire in 1876. Polished and etched face shows a Widmanstätten pattern (see text) and dark nodule of troilite (scale bar, 5 cm). *Inset:* plates of the body-centred cubic mineral kamacite occur parallel to the four octahedral planes of the face-centred cubic parental mineral, taenite. On sections cut almost parallel to the cube face of the parental taenite crystal, as in Rowton, kamacite plates appear mainly in two directions.

small number of irons containing less than 5 wt% Ni are composed of crystals of kamacite and are called *hexahedrites*, whereas irons rich in Ni (15–60 wt%) and composed mainly of taenite appear structureless to the naked eye and are called *ataxites*. Systematic chemical variations between irons serve to distinguish a number of 'chemical groups'. Contents of gallium, germanium, iridium and other elements that occur at trace levels in iron, together with nickel, distinguish thirteen well-defined groups of irons and form the basis of a modern chemical classification.[2] The chemical groups of irons are designated by roman numerals and letters (IAB, IC, IIAB, IIC, IID, IIE, IIF, IIIAB, IIICD, IIIE, IIIF, IVA, IVB).

Since their formation, chondritic meteorites have not undergone melting and differentiation. Chondrites are essentially aggregates of materials retaining a physical and chemical record of some of the earliest events in the history of the Solar System. In contrast, iron, stony-iron and achondritic stony meteorites mainly represent the products of extensive melting, differentiation and mixing.

Although the genetic relationships between various types of meteorite remain uncertain, the differentiated meteorites may have formed in a number of small parent planets by the melting of chondrite-like materials and the gravitational separation of molten metal from silicates. The irons presumably represent solidified and slowly cooled segregations (cores) of molten metal, and the achondrites more or less reworked samples of the predominantly silicate residue. The stony-irons may have formed by the mechanical mixing of both solid and liquid metal and silicate at various depths within their parent planets. The pallasites (Figure 7.5), consisting of metallic iron-nickel and olivine, may have formed at 'mantle-core' regions deep within differentiated planetary bodies, whereas mesosiderites (Figure 7.6) represent mixtures of metal and achondritic stony materials of diverse origins.

Figure 7.5 Stony-iron (pallasite) found in 1931 at Springwater, Canada. Dark, rounded crystals of olivine are enclosed in a network of iron–nickel metal.

Figure 7.6 Polished slice of the stony-iron (mesosiderite) found in 1857 at Mincy, Missouri, USA. Dark rounded to angular fragments of achondritic stony materials are mixed with nuggets and fine particles of metallic iron–nickel.

Figure 7.7 'Pseudo-meteorite' – glassy industrial slag. Note the abundant bubbles, or 'vesicles', which distinguish this man-made material from genuine meteorites (specimen approximately 5 cm across).

7.3 A guide to identifying meteorites

Records show that most of the meteorites that are recovered soon after their fall land within a few tens of metres of an observer. Attention is often drawn to the event either by the 'rushing' or 'whistling' noises that often accompany such a high velocity projectile, or the sound of an object hitting the ground. Witnesses are usually left in no doubt that something unusual has happened.

In most parts of the world the chances of finding a meteorite are remote. Nevertheless, curators are frequently asked by members of the public to identify suspected meteorites. Invariably they prove to be samples of terrestrial material not usually associated with the localities in which they occur, and so appear 'exotic' to the finder. In some cases the finder may have witnessed a fireball or meteor and linked this to the finding of unusual material, even though the two events are unrelated.

There are a number of natural and man-made materials that are commonly mistaken for meteorites. Perhaps the most abundant and widespread 'pseudo-meteorites' are near-spherical nodules of the mineral pyrite (iron disulphide, FeS_2) that formed as concretions in various sedimentary rocks, notably the Chalk of Southern England. Pyrite nodules often survive after the rock in which they formed has eroded away. Despite their metallic appearance, pyrite nodules are richer in sulphur than in iron, and give off a smell of burning sulphur when struck with a hammer. Weathered nodules commonly have a brown 'crust' of hydrated iron oxides such as goethite (FeOOH), but on freshly broken surfaces they display long, radiating brassy-yellow crystals, often terminating in crystal faces.

Other common 'pseudo-meteorites' include a variety of rock types that were transported during the last ice age (glacial erratics), the waste products ('slags') from the iron and steel industry, and the balls of cast-iron used to crush ore in ball-mills.

Industrial slags, which are often the most convincing of pseudo-meteorites, are composed of a variety of materials, such as cast-iron, glass and silicates, and often contain bubbles, or 'vesicles', that formed as the result of the expulsion of gas during solidification (Figure 7.7). In contrast, genuine meteorites are almost never vesicular.

What then are the common features of genuine meteorites that set them aside from terrestrial materials? Below are some general guidelines to aid the identification of meteorites.

Fresh meteorites possess matt, or shiny black fusion crusts about 1 mm thick, which formed during atmospheric passage. Some fusion-crusted surfaces display 'flow lines', where 'rivulets' of melted material have flowed along the surface of the meteorite. However, because of the diversity of meteorite types, fusion crusts can vary considerably in character and appearance, and identification of weathered or rare types of meteorite often requires an experienced eye. Exceptionally, in rare types of meteorite that are iron-poor, the crust may be creamy white (see Figure 7.3).

Metallic iron containing nickel is present to some extent in most meteorites, making them generally heavy objects. Meteoritic metal usually contains very little carbon (<0.1 wt%) and can therefore be distinguished from industrial cast-iron, which lacks nickel and is composed predominantly of iron and carbon. Commercial steels may contain nickel, but may also contain substantial amounts of manganese (1 wt%), chromium (up to 18 wt%) and tungsten carbide, which meteoritic metal does not.

Finally, the most abundant types of meteorite that are observed to fall, the chondrites, are characterized by chondrules.

7.4 Meteorite falls and finds – nomenclature, documentation and statistics

Although meteorites can be grouped into a number of taxonomic categories, each distinct meteorite fall is unique. The recording of information on individual recoveries forms an important part of their curation.

Since the birth of the science of meteoritics in the early nineteenth century, the scientists concerned have recognized that meteorites need to be allocated labels to distinguish one from another, independently of any mineralogical or chemical classification. From the earliest recoveries to the present day, meteorites have been named after the geographical location close to which they are observed to fall or are subsequently found.

The definitive list of all well-documented meteorites, *The Catalogue of Meteorites*[3] published by the Natural History Museum is recognized world-wide as the standard work of reference. The entry for each meteorite in the *Catalogue* gives name, locality (with coordinates), classification, and date of fall or retrieval, followed by a brief description of the circumstances associated with the fall or find. In addition, chemical data relevant to the meteorite's classification are quoted and referenced, and a list of the main repositories of material throughout the world is also given. Approximately 80% of all meteorites known to science are held in seven major collections (Table 7.4).

The discovery of large numbers of meteorites in uninhabited areas of the world with few geographical names presents difficulties for the established system of meteorite nomenclature. In the case of Antarctic meteorites the problem of labelling has been overcome by the use of a numbering system. For example, Yamato-74115 is an ordinary chondrite found in the Yamato Mountains region of Antarctica. The number indicates that the meteorite was the 115th to have been collected in that area in the 1974–5 field season.

Table 7.4 Repositories of major collections of meteorites throughout the world

Institution	Representation	Approximate numbers of individual meteorites represented
British Museum (Natural History), London, UK	World-wide, with many main masses of rare types of meteorite	1,435
Smithsonian Institution, Washington, USA	World-wide, including many recent finds from Antarctica	~1,500
Naturhistorisches Museum, Vienna, Austria	World-wide, including many historic European falls and finds	~700
American Museum of Natural History, New York, USA	World-wide, with a number of large masses	~900
Field Museum of Natural History, Chicago, USA	World-wide	~900
Institute of Polar Research, Tokyo, Japan	Antarctic meteorite finds	unknown
National Aeronautics and Space Administration (NASA), Houston, Texas	Antarctic meteorite finds	unknown

Guidelines governing the allocation of names to meteorites have been set down by the Committee on Meteorite Nomenclature under the auspices of the Meteoritical Society. This international society, founded in 1933, has both amateur and professional members devoted to all aspects of meteorite research. Names, or numbering systems, of all newly recovered meteorites have first to be submitted to the Nomenclature Committee for approval before they are announced in the *Meteoritical Bulletin* section of the Society's journal *Meteoritics*. Subsequently, full details of new meteorites are incorporated into the *Catalogue of Meteorites*. Inevitably there is a considerable time lag between the fall or discovery of a meteorite and incorporation of information into a new edition of the *Catalogue of Meteorites*. In the case of Antarctic meteorites up-to-date information on new discoveries by US scientists can be obtained from the *Antarctic Meteorite Newsletter* – a periodical issued through NASA Johnson Space Centre in Houston, Texas; Japanese discoveries have been documented in the *Photographic Catalogue of The Antarctic Meteorites*, published by the National Institute of Polar Research in Tokyo.

Meticulous documentation of meteorite recoveries has allowed statistical studies of the temporal and spatial distribution of falls,[4] and the recognition of meteorite 'pairs'. Paired falls are those pairs or groups of meteorites for which it has been suggested, on the basis of geographical proximity and classification, possibly or probably belong to a single fall. When two or more meteorites, found at different times and allocated separate names, are proved conclusively to be from the same fall, they are said to be 'synonymous', and the name of the meteorite that was recovered first, or has the largest mass, takes precedence.

Inconsistencies in the literature on meteorites have provided an additional source of synonyms. In most cases one name may have been used at different times for several distinct falls. To avoid confusion, special attention is given to synonyms in the *Catalogue of Meteorites*, and the accepted name of one meteorite may be recorded as a synonym of two or more others. As an illustration, 'Washington County', an iron meteorite found in 1927 in Colorado, is also a synonym for the ordinary chondrite that fell in 1890 at 'Farmington' and the iron meteorite found in 1858 at 'Trenton' in identically named counties of Kansas and Wisconsin, USA, respectively.

Well-documented meteorite falls provide the best available measure of the relative abundances of the different types of meteorite that survive their fall to Earth. The fall frequency (Table 7.3) is the number of each type of meteorite observed to fall expressed as a percentage of the total number of recorded falls.

Excluding stony-irons and irons, which are the rarest types seen to fall, the relative numbers (falls and finds) of chondrites (enstatite carbonaceous and ordinary) and achondrites that are represented in collections correspond well with those predicted by their fall frequency. In contrast, irons and, to a lesser extent, stony-irons appear to be over-represented in collections. The reason for the disproportionate number of iron and stony-iron meteorite finds is that, by virtue of their exotic nature, the strongly metallic meteorites are most easily recognizable. Indeed records show that ancient civilizations found, utilized (Figure 7.8) and even revered meteoritic iron, even though, in many cases, they had no inkling of their origin.

Astronomical studies of the absorption spectra of asteroidal bodies indicate that carbonaceous matter may be the most abundant among the potential source materials of meteorites. Carbonaceous meteorites are generally friable, weak materials, and the destructive effects of passage through the Earth's atmosphere, combined with the rigours of the terrestrial environment, are probably responsible for the small numbers of this type of meteorite in collections.

The aim of statistical studies is to obtain a reliable value for the influx of meteorite falls on the Earth. Statistics based on witnessed falls and chance recoveries are fraught with difficulties, and generally rely on *ad hoc* assumptions about how the distribution of human populations around the world and their varying levels of education affect the probability of meteorite recognition.[5]

Major networks of cameras covering large areas (10^6 km²) of the earth's surface provide more reliable estimates of the current annual influx of meteorites. In recent years, as the result of advances in the understanding of the luminous phenomena associated with the fall of a meteorite, it has become possible to identify those fireballs with the potential to yield meteorites, even if none are subsequently recovered. Recent estimates indicate that up to 19,000 meteorites may fall annually over the entire surface of the Earth, including some 5,000 falls depositing meteorite masses in the range 0.1–10 kg on the land surface.[6] However, camera techniques are not yet sufficiently sophisticated to distinguish with confidence meteorite types from observations of fireballs.

7.5 Conservation and storage of meteorites

From the moment a meteorite enters the Earth's atmosphere it is subject to contamination from the terrestrial environment. Since all meteorites have come into contact with the atmosphere before they are

Figure 7.8 Two bone-handled knives made by Greenland Eskimos. The cutting edges are fashioned out of metal from iron meteorites found at Cape York, in north-west Greenland.

recovered, it is impossible to eliminate terrestrial contamination completely, and meteorite 'falls' and 'finds' present quite different conservational problems.

In the case of freshly fallen and rapidly recovered meteorites the overriding aim of preservation is to minimize further contamination. Paradoxically, the fusion crust, which is the first and most obvious effect of terrestrial contamination, can often act as an effective temporary seal against contamination of the pristine interior of the meteorite. Although it is best to avoid direct handling of fresh meteorite samples, this may be done, if absolutely necessary, by means of the crust.

Effective preservation of meteorites depends largely on financial considerations. Meteorites recovered from Antarctica are transported in their frozen state and are curated under conditions similar to those afforded lunar samples returned to Earth by the Apollo space missions. While they approach the ideal, the sterile storage facilities constructed by NASA Johnson Space Centre in Houston are beyond the means of most institutions and museums, and are impractical for most meteorites. Cheaper and fairly effective preservation of meteorites can be achieved through an understanding of some of the common types of contamination, their effects on meteoritic materials, and the ways in which problems can be combated.

7.5.1 The rust problem

The most potent form of contamination is corrosion, and the most damaging type of corrosion is that brought about by moisture and air. The presence of abundant metallic iron in most types of meteorites renders them susceptible to rusting, the problem usually increasing with the abundance of metal. Rusting is an electrochemical reaction of metal surfaces with water vapour and oxygen in the atmosphere, resulting in the formation of oxyhydroxides of iron such as goethite (α-FeOOH), lepidocrocite (γ-FeOOH), akaganéite (β-FeOOH), and other corrosion products such as maghemite (γ-Fe$_2$O$_3$). The increase in volume caused by the conversion of metal to hydrated oxide can often disrupt a meteorite along planes of weakness (Figure 7.9).

Although the cause of rusting is known, and the devastating effect of the reaction on meteorites is often all too apparent, why some meteorites rust more readily than others has been, until recently, incompletely understood. For example, many iron meteorite finds that have been stored in collections for a century or more show few signs of corrosion, whereas others, that were well preserved when found, have rusted badly on removal from the soil.

Most meteorites contain trace amounts of chlorine. In some meteorites analysts have reported up to several per cent Cl$_2$ and suggested that a chlorine mineral, lawrencite (FeCl$_2$), is present. The occurrence of primary lawrencite has never been substantiated adequately, and it is significant that high concentrations of chlorine are found only in weathered finds that have suffered prolonged exposure to ground waters. It is now generally accepted that lawrencite is not a primary meteoritic mineral and that virtually all chlorine in iron

Figure 7.9 Rusty fragment of the iron meteorite (hexahedrite) found in 1940 at Smithonia, Georgia, USA. Extreme corrosion (rusting) has caused the sample to split along cubic planes in the mineral kamacite.

meteorites is of terrestrial origin.[2] However, irrespective of their origin, ferric and ferrous salts can greatly accelerate rusting in meteorites.

Recent research on Antarctic meteorites by Buchwald and Clarke[7] has greatly enhanced our knowledge of the rusting process. They have shown that the major terrestrial corrosion products of meteoritic metal in the Antarctic environment are akaganéite, an otherwise rare terrestrial mineral, and goethite. Significantly, they found that akaganéite containing up to 5% by weight chlorine forms as a replacement for metal at the active corrosion front. With time and distance from the corrosion front, akaganéite loses its chlorine and decomposes *in situ* to intergrowths of maghemite and goethite, that form crusts of largely inactive oxides. The significant properties of the crystal structure of akaganéite are its ability to incorporate Cl⁻ ions attracted electrolytically to the corrosion front, and its small crystal size, which greatly increases its water-absorption capacity. Buchwald and Clarke[7] suggest that, under moist conditions, Cl⁻ ions incorporated in the akaganéite structure are released to return to the akaganéite/metal interface where they depassivate the metal and reinitiate corrosion. Attacking akaganéite may again incorporate Cl, providing a

source of corrosive agent behind the reaction front. The work of Buchwald and Clarke[7] demonstrates that terrestrial chlorine may be derived and accumulated in meteorites even in environments where it is present in very low concentrations and, because of the electrochemical nature of the corrosion reaction, very small amounts of chlorine can lead to the destruction of large irons.

There are a number of procedures in which caustic solutions may be used to remove or neutralize ferric salts, or pure ethanol (ethyl alcohol) is used to flush chlorine from the corrosion products. However, meteorite finds are often deeply fissured, and leaching is often risky. Unless it is complete and the specimen thoroughly dried, the process may produce more corrosion than that it was intended to treat. More importantly, care has to be taken that the leaching solution used will not react with the primary mineral assemblages in the meteorite. Routine chemical treatment of 'rusty' meteorites is not recommended, and should be reserved only for those cases where, unless drastic steps are taken, there is a danger that the material may be completely destroyed by corrosion.

Both oxygen and moisture are necessary for rusting to occur, and the exclusion of either one or

Figure 7.10 An individual from the shower of carbonaceous chondrites that fell in 1864 at Orgueil, France. Slow decomposition of epsomite (white) has caused the fusion crust to flake away and the meteorite to disintegrate.

the other prevents the reaction. Since the exclusion of oxygen is not practicable, protection against corrosion may be achieved by preventing contact with moisture. One of the most effective methods of protection is the incorporation of a powerful moisture-absorbing material with the meteorite in sealed containers. Heavy gauge polythene bags provide watertight, durable and economically replaceable packaging for meteorite samples. Those meteorites containing a high percentage of metallic iron are enclosed in two individually sealed polythene bags (one inside the other). Into the inner bag, along with the meteorite, is placed a plastic tube plugged with cotton wool, or an open-weave canvas bag filled with self-indicating silica gel or other moisture absorber. Meteorites prone to rusting should be monitored closely, and silica gel and packaging changed at regular intervals.

7.5.2 Deliquescence and efflorescence

More destructive and more difficult to control than rusting is the effect of the atmosphere on water-soluble minerals, and on those minerals that deliquesce or effloresce (see pp. 33–40). Efflorescence is the loss of water of crystallization, which may cause a mineral to crumble to a powder, whereas deliquescence is the absorption of water by a mineral, effectively dissolving it.

Many carbonaceous chondrites contain appreciable amounts of the mineral epsomite ($MgSO_4.7H_2O$), which dissolves readily in water. Epsomite effloresces, and decomposition can cause the meteorite in which the mineral occurs to disintegrate slowly (Figure 7.10). Since carbonaceous chondrites can contain up to 20 wt% water, much in the form of hydrous sulphates, the possibility of self-destruction with time is impossible to combat. An example is the epsomite-rich carbonaceous chondrite that fell at Alais in France in 1806. Of the two individuals, one weighing 2 kg and the other 4 kg, which were recovered, only fragmental material survives to the present day.

A further example is the rare mineral oldhamite (CaS), which occurs as a minor constituent of meteorites with highly reduced mineral assemblages such as enstatite chondrites and achondrites (Figure 7.3). While calcium sulphide is fairly insoluble in water, exposure to air and moisture causes decomposition and the hydrolyzed products dissolve rapidly.

Fortunately the more common sulphides that occur in meteorites as primary minerals, such as troilite (FeS), daubréelite ($FeCr_2S_4$) and pentlandite

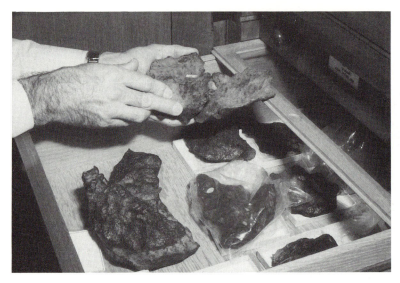

Figure 7.11 Storage of small samples of meteorites at the British Museum (Natural History), London.

(FeNi$_9$S$_8$), are relatively stable. In some deeply weathered meteorite finds troilite has reacted with nickel liberated by the oxidation of metallic iron–nickel to form secondary pentlandite, but there is no evidence that this occurs during prolonged storage in collections.

7.5.3 Storage

The need to prevent contamination or corrosion requires that meteorites are stored at low to moderate temperatures, under dust-free and dry conditions, as effectively sealed from the atmosphere as practicable. Most meteorites are composed of dense materials, and weight is also an important consideration. Well-made drawers with tightly fitting, sliding glass tops, in strongly built cabinets, provide good storage for small to medium-sized meteorite samples up to a few kilograms in weight (Figure 7.11).

Storing large or heavy meteorites is difficult. Specimens of a few tens of kilograms in weight can be wrapped in polythene and stored in strong cupboards, but meteorites weighing several hundred kilograms or more present logistical problems that may be compounded if special curatorial attention is required. A splendid example is the 3.5 ton mass of the 'Cranbourne' meteorite in the display of meteorites at the Natural History Museum (Figure 7.12). This iron meteorite, found in Victoria, Australia, in 1854, rusted badly on removal to the temperate British climate in 1863. During storage in the Museum's collection, some 25 kilograms of rusty flakes have fallen from the surface of the mass. In 1936 the meteorite was enclosed in a sealed glass case under a constantly changing atmosphere of dried nitrogen. This treatment, while not eliminating rusting completely, dramatically reduced the rate of disintegration of the mass.

7.6 Curatorial guidelines

Important scientific information that may be gained from a study of meteorites can easily be prejudiced by thoughtless curation. In general, curatorial practices described elsewhere in this volume relating to other mineralogical and petrological collections apply equally to meteorites. Practices deviate only where there is a risk of contaminating pristine meteorite samples. For example, paints and glues commonly used to affix numbers to specimens are potential contaminants. If necessary, numbers are affixed to the fusion-crusted surfaces of meteorites. Specimens of rare and important types of meteorite, such as carbonaceous chondrites, should not be numbered physically. Keeping an accurate and up-to-date record of the weights of specimens ensures that they can be identified independently of any fixed labels or numbers.

Similar principles apply when meteorites are sampled for research or prepared for display. To avoid contamination from saws and cutting lubricants, fresh stony meteorites should be broken. This has the added advantage that no material is lost, but breaking small samples from meteorites requires some skill. Weathered stony meteorite finds that are already heavily contaminated may be cut, but if

Figure 7.12 Apparatus supplying dried nitrogen to the case (above) containing the Cranbourne iron meteorite. The nitrogen is dried by bubbling through sulphuric acid.

water is used as a cutting lubricant and coolant, care should be taken to ensure that the meteorite does not contain soluble minerals, or that cutting does not trigger corrosion.

Because of the ductility of iron–nickel metal, dividing iron meteorites unavoidably causes some mechanical damage. Breaking iron meteorites is neither easy nor desirable, and they have therefore to be cut. Saws commonly used to cut irons include ordinary or powered hacksaws and more specialized, industrial high-speed metallurgical saws. Structural alteration to the metallic minerals in irons can be induced at temperatures of only a few hundred degrees centigrade, and so the generation of frictional heat during cutting must be avoided.

Saws utilizing a rapidly rotating, continuous loop of steel wire fed with a slurry of carborundum and water, similar to those used by stonemasons, have proved ideal tools for cutting meteorites of all sizes. This method of cutting iron meteorites, pioneered by Danish scientists on a 20-tonne mass of the 'Cape York' iron,[8] inflicts minimal mechanical damage, generates very little heat and results in a small loss of material. Wire-saws cut very slowly. In the Danish example, approximately 390 hours were taken to remove a slice measuring 180×130×5 cm from the 'Cape York' meteorite. Smaller, bench-top versions of 'wire-saws' are commercially available and can be used to cut small meteorites accurately, and with minimum wastage of material.

7.7 Preparation of iron meteorites for display

Large slices of octahedrite iron meteorites that have been etched to reveal their Widmanstätten patterns feature as the centre-pieces of almost every museum display of meteorites (Figure 7.13). To reveal the structure, even in small specimens, requires many hours of careful preparation.

To remove the damage caused by cutting, the surface to be etched must be ground to a smooth, perfectly flat finish, using successively finer grades of carborundum (down to about 600 mesh). Small samples may be polished further by using diamond pastes. Between each grinding or polishing stage the specimen is washed clean. When a suitable surface has been prepared, the specimen is 'de-greased' by swabbing with a weak soap solution, then rinsed clean and thoroughly dried in a stream of warm air.

A freshly prepared solution of concentrated nitric acid and grain alcohol (nital) – normally 5 ml HNO_3 in 95 ml C_2H_5OH – is poured on to the polished surface so that the entire surface is quickly covered. By means of a cotton wool swab or wad of soft cloth

Figure 7.13 Polished and etched slice of an octahedrite iron meteorite showing the Widmanstätten pattern. Kamacite bands (light) appear in three directions, bounding dark areas of the parental mineral taenite. Kamacite bands in the fourth octahedral direction, which is roughly parallel to the plane of the slice, appear as irregular 'blooms'.

the etching solution is scrubbed over the entire surface. It is important that all the surface receives about the same amount of scrubbing.

Within a few minutes the Widmanstätten structure begins to emerge. Sometimes one application of the etching solution is sufficient, but if the structure is not sharply defined, a second or third application is made. Depending on the strength of the etching solution used, the structure may be developed fully in less than 15 minutes but may take 45 minutes. (Note that the etching solution, nital, is a potentially explosive mixture, and should not be stored.)

When etching is complete, the etchant is washed off with alcohol. The specimen should be washed again in alcohol, and this is removed from the etched surface with a soft, clean brush. Next the specimen is doused in a stream of water for two to three minutes, and with continuing brushing. The etched surface is then repeatedly washed (at least twice) in alcohol to remove the water, and finally set up in front of a fan drier in such a way that the surface will drain and dry in the stream of warm air.

Local tarnished areas on the surface can sometimes be removed by lightly scrubbing with a clean, soft rubber eraser. At all stages of the preparation, care must be taken not to touch the etched surface, because fingerprints will show up as dark lines (see pp. 52–53).

It is often forgotten that the role of a curator extends further than that of a mere custodian of rare material. Frequently it is the curator's unenviable task to arbitrate over whether, or on what terms, material should be sacrificed for destructive research. In most cases there are no hard and fast rules, and requests for material have to be judged on their merits. Decisions are usually governed by the need to obtain maximum scientific gain from minimum loss of material. This requires curators to stay abreast of scientific and technical developments in fields that are often well outside their own areas of specialization. In multi-disciplinary subjects such as meteoritics this, in itself, is no easy task.

References

1 GRAHAM, A.L. and ANNEXSTAD, J.O., 'Antarctic meteorites', *Antarctic Science*, 1 (1), pp. 3–14 (1989)

2 BUCHWALD, V.F., *Iron meteorites*, 3 Volumes, University of California Press, Berkeley (1975)

3 GRAHAM, A.L., BEVAN, A.W.R. and HUTCHISON, R., *The Catalogue of Meteorites*, 4th edition, British Museum, Natural History (1985)

4 HUGHES, D.W., 'Meteorite falls and finds: some statistics', *Meteoritics*, 16, pp. 269-81 (1981)

5 WICKMAN, F.E. and PALMER, C.D., 'A study of meteorite falls, *Proceedings of the Indian Academy of Sciences*, section A, 88, pp. 247-72 (1979)

6 HALLIDAY, I., BLACKWELL, A.T. and GRIFFIN, A.A., 'The frequency of meteorite falls on Earth', *Science*, 223, pp. 1405-7 (1984)

7 BUCHWALD, V.F. and CLARKE, R.S. JR, 'Corrosion of Fe-Ni alloys by Cl-containing akaganéite (β-FeOOH): The Antarctic meteorite case', *American Mineralogist*, 74, pp. 656-67 (1989)

8 BUCHWALD, V.F., 'A new cutting technique for meteoritic irons', *Meteoritics*, 6, pp. 27-31 (1971)

Further reading

DODD, R.T., *Meteorites - a petrologic-chemical synthesis* Cambridge University Press, Cambridge (1981)

DODD, R.T., *Thunderstones and Shooting Stars - The Meaning of Meteorites* Harvard University Press, Cambridge, Mass. and London (1986)

HUTCHISON, R., *The Search for Our Beginning* British Museum of Natural History and Oxford University Press (1983)

MCSWEEN, H.Y., *Meteorites and their Parent Planets* Cambridge University Press, Cambridge (1987)

8

The lunar sample collection

Charles Meyer, Jr

The moon has been directly sampled in nine separate places. Additional samples have been received indirectly from other sites via exotic rocks found at the Apollo sites and via nine meteorites from the Moon. Lunar samples consist of basalts from the mare regions, breccias from the ejecta of the giant impact craters, and anorthositic materials from the highlands. Numerous soil samples and core tubes provide the opportunity to study the lunar regolith. Lunar samples consist mostly of calcic plagioclase, olivine, orthopyroxene, ilmenite and glass. Only three new minerals were found.

The collection of samples from the Apollo missions are carefully curated at the Johnson Space Centre in Houston, Texas. They are stored in nitrogen cabinets and cut with a bandsaw as needed. Much of the collection is a working collection studied by scientists world-wide. Some of the samples are used for educational purposes.

Twelve men walked on the surface of the Moon and carefully collected 2,196 documented samples of soils and rocks from the lunar regolith during 80 hours of exploration.[1] After careful preliminary examination, a large number of splits of these samples were distributed to scientists for detailed mineralogical, chemical and physical analysis, and a few samples have also been loaned for public displays in museums and presidential libraries. Most of the lunar sample collection is still stored in an ultra-clean laboratory at the Johnson Space Center in Houston, Texas.

The lunar samples were collected by the astronauts at great personal risk. Altogether, 24 astronauts travelled to the Moon during nine trips (three astronauts went twice). The explosion of the fuel cell during the Apollo 13 mission emphasized the dangers of space travel; there was also the added danger of radiation from potentially fatal solar flares. Today, watching the films of the astronauts working on the lunar surface, it is hard to realize that they were working in a complete vacuum. Some of the tasks, such as the withdrawal of the Apollo 15 core, were quite difficult. On the last three missions there was the potential need for a long walk back to the Lunar Module in case the rover broke down. For geoscientists, the legacy of Apollo is to try to solve the puzzles presented by the samples that these men diligently collected during this extraordinary adventure.

Samples from the early missions were returned in sealed rock boxes; however, other samples were exposed to the atmosphere of the Lunar Module, Command Module and even (briefly) to the atmosphere of the equatorial Pacific Ocean. Once they reached Houston, they were stored in pure nitrogen. Originally the lunar samples were quarantined in the Lunar Receiving Laboratory to make sure that no extraterrestrial life forms were present in the samples. During quarantine the samples were given a preliminary examination and catalogued, and the rocks were orientated by means of artificial lighting to match the shadow patterns on the photographs taken by the astronauts. The results of the preliminary examinations were used to plan the distribution of samples to scientists for more detailed studies, which could best be done in their highly specialized laboratories. No life forms, organic molecules or water were found in the lunar samples, so the quarantine was discontinued.

After quarantine, the lunar samples received the most extensive and comprehensive analysis of any geological collection ever, by an international team of principal investigators chosen by peer review. The results of these studies are reported annually at the Lunar and Planetary Science Conferences, and the *Proceedings*[2] of these conferences are the best way to access the literature. Analysis of the samples showed that there was a very diverse set of lunar rocks, indicating early differentiation of the outer portion of the Moon. Plagioclase flotation from a global magma ocean apparently formed the early lunar crust. Trace-element-enriched residual liquids initially crystallized 4,400 million years ago; later, about 3,900 million years ago, the giant basins formed during extensive bombardment of the lunar crust; and remelting of the lunar interior produced iron-enriched basalts from about 4,000 to 3,200 million years ago. Continued meteorite bombardment of the lunar surface formed a thick layer of debris called regolith.

8.1 Lunar mineralogy

The mineralogy of lunar samples is rather simple, with only a few major minerals (plagioclase, pyroxene, olivine and ilmenite).[3,4] The rocks formed in a completely dry and very reducing environment. Most of the iron is in a plus two oxidation state with a minor amount of metallic iron. The grain boundaries between minerals are remarkably distinct, and there are no alteration products. Glass is present in the mesostasis of the igneous rocks. Minerals that might have been added by meteorites have all been melted or vaporized by impact.

There are a few unique features in lunar rocks (Table 8.1): lunar plagioclase is almost pure anorthite (calcium feldspar); maskelynite (shocked plagioclase) is common; and feldspars with ternary (Ca, Na, K) composition were found in rare lunar felsite particles. Lunar pyroxenes have a wide range of composition, and those from the plutonic fragments have interesting exsolution features. New minerals that were found included armalcolite, tranquillityite and pyroxferroite. Quenched, iron-rich and silica-rich immiscible liquids were found in the mesostasis of mare basalts. Rocks that were exposed to the micrometeorite environment have a patina of glass-lined microcraters and glass splashes, large and small. Surface coatings of ZnS were found on some volcanic glasses. Akaganéite (FeOOH) was found on the surface of one Apollo 16 breccia.

Lunar minerals do not react appreciably with the earth's atmosphere. The fine-grain soil gains weight only slowly on an analytical balance and the polished thin sections do not tarnish. The only identifiable problem is slow oxidation of iron grains, but the classic problem of oxidation of iron does not seem to be the problem that it is with some meteorites. In meteorites of terrestrial origin, oxidation is probably caused by the catalytic action of lawrencite acquired in the atmosphere or later. Lawrencite does not appear to occur in lunar rocks.

8.2 Basic descriptions

The Apollo Collection is made up of 382 kilograms of rocks and soils (Table 8.2). In addition to the Apollo Collection, three cores were returned by the USSR using unmanned spacecraft. In addition nine meteorites have now been identified as having a lunar origin. Each mission returned samples unique to that site, as well as exotic samples from other places on the Moon. There are about twice as many breccias as basalts, and there are numerous samples of regolith material. Almost all the samples have been described briefly and many have been studied in great detail.

8.2.1 Basalts

An extensive collection of very fresh, yet very old, basalt samples were returned from the mare surfaces of the Moon. There are 134 samples of basalt greater than 40 grams, 42 greater than 500 grams, 24 greater than 1 kilogram, 11 greater than two kilograms and the largest (15555) is 9.6 kilograms. They have a wide variety of textures, ranging from variolitic to subophitic to equigranular. Most are fine-grained, with an average grain size about 0.5 mm, but some have phenocrysts over one cm. Many have a high percentage of opaque minerals and some have metallic iron grains. Some are very vesicular, with interconnecting vugs and vesicles (Figure 8.1), but the

Table 8.1 Lunar minerals

Major minerals	Rough formula
Plagioclase – mainly anorthite	$CaAl_2Si_2O_8$
Pyroxene	
ortho	$(Mg,Fe)SiO_3$
clino – pigeonite	$(Ca,Mg,Fe)SiO_3$
Olivine	$(Mg,Fe)_2SiO_4$
Ilmenite	$FeTiO_3$
Minor phases	
Armalcolite	$(Mg,Fe)(Ti, Zr)_2O_5$
Tranquillityite	$Fe_8(Zr,Y)_2Ti_3Si_3O_{24}$
Zirkelite – zirconolite	$(Ca,Fe)(Zr,Y,Ti)_2O_7$
Chromite – ulvospinel	$FeCr_2O_4$
Iron	Fe,Ni,Co
Troilite	FeS
Spinel – pleonaste	$MgAl_2O_4$
Zircon	$(Zr,Hf)SiO_4$
Baddeleyite	ZrO_2
Rutile	TiO_2
Apatite	$Ca_5(PO_4)(F,Cl)$
Whitlockite	$Ca_3(PO_4)_2$
Silica	SiO_2
Ternary feldspar	$(Ca,Na,K)AlSi_3O_8$
Ba-sanadine	$(K,Ba)AlSi_3O_8$
Pyroxferroite	$(Fe,Ca)SiO_3$
Symplectite: in 76535, 72415	Cr spinel, 2 pyrox
Akaganéite – on surface of 66095	FeOOH
Cordierite, clasts in 15445, 73263, 72435	$Mg_2Al_4Si_5O_{18}$

Table 8.2 Lunar sample inventory (circa 1988)
30 educational thin-section packages (12 each)
99 lunar educational disks (6 each)
9,000 petrographic thin sections

USA	Weight	Number
By location		
Brooks AFB	50 kg	343
Pristine vault	280	20519
Returned vault	24	33548
Principal investigators	8	12082
PAO	8	1648
Gifts	.2	482
Consumed during anal.	12	15344
Total	382 kg.	84000

By mission			*EVA*	
Apollo 11	21.5 kg	58	2.2 hrs	0.5 km
Apollo 12	34.4	69	7.6	2.0
Apollo 14	42.3	227	9.2	3.4
Apollo 15	77.3	370	18.3	23
Apollo 16	95.7	731	20.1	20.7
Apollo 17	110.5	741	22.0	31.6
Total	382 kg	2196	80 hrs	81 km

By type		
Soils	80 kg.	167
Breccias	133	79 over 300 grams
Basalts	80	134 over 40 grams
Cores	20	24 holes
Other	69	(mostly small breccias)

Total length of cores is 15 metres (52 segments).

Lunar meteorites	
ALHA81005	.031 kg.
Y791197	.052
Y82192	.700
Y793274	.009
MAC88104,5	.700
EET87521	.031
Asuka 31	.442
Y793169	.006
Calcalong Creek	.018

USSR Cores		
Luna 16	.101 kg.	35 cm.
Luna 20	.050	27
Luna 24	.170	160

composition of the gas phase has never been determined. All lunar samples are very reduced, with iron in the plus two oxidation state. These mare basalt samples represent pristine (uncontaminated by meteorite) volcanic liquids, presumably from deep in the lunar interior.[5] Mare basalts are very iron-rich ($FeO = 20\%$) compared with most terrestrial basalts. Some are also very titanium-rich ($TiO_2 = 13\%$), and some are very magnesium-rich ($MgO = 20\%$). Mare basalts range in age from 3,200 to 4,000 million years. Clasts of mare basalt are also found in some lunar breccias.

Another variety of lunar basalt (termed KREEP basalt, because of its high trace-element composition) was found in abundance at the Apollo 14 and 15 sites.[6] The major element composition of KREEP basalt indicates that plagioclase was in the source region, but only small fragments of this variety of lunar basalt have been found to be free of meteoritic contamination. Measured ages of pristine KREEP basalts (15382, 15386) were about 3,900 million years.

8.2.2 Breccias

Most lunar breccias are the lithified aggregates of clastic debris and melt generated by meteoritic bombardment in the ancient lunar highlands about 3,900–4,000 million years ago. There are 59 lunar breccias larger than 500 grams, 39 greater than one kilogram and 19 greater than two kilograms. Many of the breccia samples are ejecta from the giant

Figure 8.1 Lunar basalt 15016 (924 grams) is the most vesicular mare basalt, with about 50 per cent vesicles. The vesicles are about 3 mm in size and interconnecting. The gas phase that produced the vesicles is unknown. The mineralogy is 52 per cent clinopyroxene, 25 per cent plagioclase, 18 per cent olivine and 5 per cent ilmenite. The texture of this mare basalt is ophitic, and the average grain size is 0.3 mm. The scale is 1 cm (NASA S71–46632).

Figure 8.2 Lunar breccia 14306 (583 grams), from the Fra Mauro Formation–ejecta from the giant Imbrium Basin. This sawed surface of 14306 shows that there are several generations of breccia included as clasts. Most of the mineral fragments are orthopyroxene or plagioclase. A few clasts of mare basalt are included. The scale is 1 cm (NASA S77–22103).

basin-forming events.[7] Others are interpreted as melt sheets from the fallback of ejecta into large lunar craters. Some have a fragmental matrix made up of individual mineral fragments, while others have a crystalline matrix from slow cooling of initially molten matrix. A few are soil breccias containing glass beads and a component of the solar wind. Most breccias are polymict (Figure 8.2), and contain a wide variety of clasts. However, most clasts are themselves breccias. The determination of trace amounts of gold and iridium is a crucial measurement in the study of lunar breccias and their clasts,

because these elements indicate the amount of admixed meteorite component. Breccia clasts without gold or iridium are termed 'pristine', and represent pieces of the original lunar crust before meteorite bombardment. Using trace siderophile and volatile elements, some scientists have even assigned breccias to specific lunar craters.[8]

8.2.3 Plutonic fragments

An early discovery of lunar sample analysis was that the lunar highlands contained an abundance of

Table 8.3 Pristine plutonic fragments of the original lunar crust[2]

Sample	Type	Weight	Age (my)	Technique	Special feature
72415	Dunite	55 grams	4550±100	Rb/Sr	Symplectite
76535	Troctolite	155	4260±60	Sm/Nd	Symplectite
78235	Norite	400	4340±40	Sm/Nd	
77215	Norite	846	4370±70	Sm/Nd	
73255,27	Norite clast	clast	4230±50	Sm/Nd	
72255	Norite	10	4170±50	Rb/Sr	Cataclastic
76255	Norite	300			
15455	Anorthositic Norite	200	4480±120	Rb/Sr	
72435	Mafic clast				Cordierite
15445	Mafic clast				Cordierite
67435	Spinel troctolite	2			Mg spinel
15405,57	Monzodiorite clast	3	4310±30	U/Pb	Zircon
14321,1062	Granite clast	2	3960±20	U/Pb	Zircon
14303,209	Granite clast		4310±30	U/Pb	Zircon
14306,60	Sodic ferrogabbroclast		4350±30	U/Pb	Zircon
15362	Anorthosite	4			
15415	Anorthosite	269	3890±70	K/Ar	
60025	Anorthosite	1836	4440±20	Sm/Nd	
67075	Anorthosite	219			
67667	Feldspathic lhersolite	8	70	Sm/Nd	
78155	Granulite	401	4170±20	U/Pb	

anorthositic material without a significant quantity of an equivalent mafic complement. This has been interpreted to mean that the anorthositic crust of the Moon is relatively thick, otherwise the large basins would have dug up more mafic samples. Table 8.3 lists the 'pristine plutonic fragments of the original lunar crust'. Only one fragment of dunite (72415) was found; both this sample and a troctolite (76535, Figure 8.3) were found to be very old–approximately 4,400 million years. Lunar anorthosites should also be quite old, but they have proved to be difficult to date. The anorthosites have relatively high Fe/Mg ratios and are termed ferroan anorthosites.[9] Both 60025 and 15455 have been measured to be as old as 4,400 million years. A second suite of plutonic rocks (termed Mg-norites) has been recognized,[10] but it is not thought to be directly related to the ferroan anorthosite suite, because of the different trend in Fe/Mg ratios (Figure 8.4). Samples of this suite (78235 and 77215) have been measured to be roughly 4,350 million years old.

Late-stage differentiates of layered igneous intrusions or of a large-scale magma ocean include granite 14303, sodic ferrogabbro 14306 and quartz monzo-

Figure 8.3 Lunar troctolite 76535 (155 grams), one of only a few pristine plutonic fragments that escaped impact metamorphism. It contains about 58 per cent plagioclase, 37 per cent olivine and 4 per cent pyroxene. The grain size is about 5 mm and the minerals commonly meet at 120 degree angles, indicating that the rock was annealed at high temperature. The olivines include a minor amount of symplectite intergrowth of Mg-Al chromite and pyroxene. The scale is in mm for this close-up photograph (NASA S73-19601).

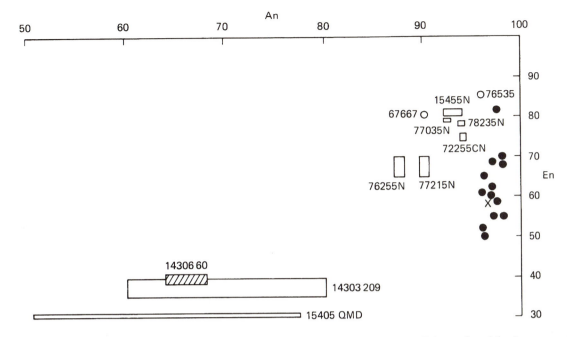

Figure 8.4 The compositions of plagioclase (An) and of co-existing low-Ca pyroxene (En) are plotted for fragments of pristine plutonic lunar samples. The X is for anorthosite 15415 and the dots are for other ferroan anorthosites from Apollo 16. The samples of lunar norites are followed by an N. Samples 15405 (quartz monzodiorite), 14306,60 (sodic ferrogabbro) and 14303,209 (granite) are late-stage differentiates of layered igneous intrusions and/or of a global lunar magma ocean.

diorite 15405. Zircons in these granitic fragments are as old as 4,350 million years. Other clasts of lunar felsite have ages as young as 3,900 million years.

8.2.4 Glass

Glass is an important component in lunar samples, and has been studied by many researchers. Glass occurs as mesostasis in basalts, as beads and agglutinates in soil and as splash on rocks. Agglutinates are fragment-laden–vesicular glasses that are made from solar-wind-enriched lunar soil by micrometeorite bombardment. In addition, there are also volcanic glasses, presumably made by ancient fire fountains on the Moon. Suspected volcanic glasses include the orange glass soil (74220), clods of green glass (15426) and numerous individual beads in other soils.[11] Some investigators have measured the compositions of hundreds of individual glass fragments and interpreted groupings in glass compositions to represent 'rock types', but these compositions may be mixtures.[12]

8.2.5 Soils

The lunar regolith is the interface of the lunar surface with the harsh space environment.[13] It is a mixture of a variety of rocks and soils, derived by meteorite bombardment of the lunar surface. The lunar soils that have been exposed for a long time have a large amount of fine material (half is less than 50 microns) and a high percentage of agglutinate glass; these soils are termed mature.[14] However, the 80 kilograms of soil samples also contain over 1 million 'coarse–fine' particles (1–10 mm), which are each large enough for scientific studies.

Since some visible rays from large lunar impact craters extend half-way around the Moon, everything on the moon should be represented in the soil samples. However, to a first approximation, the compositions of lunar soils can be successfully matched with mixtures of known lunar rock types.[15] Only one or two per cent of exotic component can be accommodated. There is also about a two per cent meteorite component in mature lunar soils. Immature soils were collected from trenches, from under rocks, where they were shielded, and from fillets around boulders. Altogether a total of 167 carefully documented soil samples were collected. There are 35 soils over 500 grams, 15 over one kilogram, four over two kilograms and the largest (14163) weighs 7.8 kilograms. A third of each soil remains unsieved, to avoid contamination and to preserve delicate features.

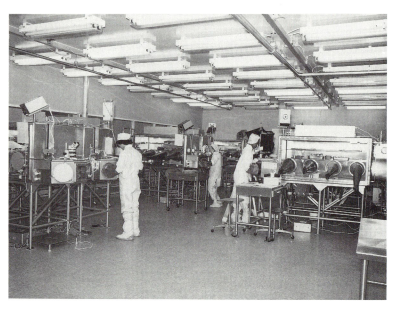

Figure 8.5 The processing laboratory of the Lunar Curatorial Facility in Houston TX contains nitrogen-filled glove boxes in which the lunar samples are viewed and described with binocular microscopes, sub-divided, weighed and packaged for distribution to scientists. The glove boxes each connect to a 'pristine' hallway, which leads to the lunar sample vault containing additional nitrogen cabinets for storage. The samples, packaged in Teflon bags and stainless-steel cans, never leave the pristine vault, hallway or glove boxes unless they are allocated for scientific study. Separate processing and storage cabinets are used for each mission to avoid cross-examination of samples from different missions. Visiting scientists often help with descriptions of samples in this laboratory (NASA S80-29341).

8.2.6 Cores

Twenty-one drive tubes were hammered into the regolith and three long drill strings were taken from the later missions.[13,16,] Some of the drive tubes were four cm in diameter, but the drill strings were only two cm in diameter. Eighty per cent of these cores have now been carefully dissected and studied, and 20% are stored for future dissection. Aliquots from different depths are stored in individual containers, and contiguous thin sections have been prepared for the entire length. The total length of the cores is 15 metres and the total weight is 20 kilograms. There are 53 separate segments of which only 6 remain unopened. The longest drill (from Apollo 17) was 2.86 metres long. The cores contain an interesting record of stratification, which is controlled primarily by small craters in the local regolith.

8.3 Curatorial laboratory

Three different buildings have housed the lunar samples since they were returned to Houston. They were initially quarantined and catalogued in the Lunar Receiving Laboratory. From 1972–1979 they were processed in clean laboratories in Building 31. Finally, in 1979, a permanent Lunar Curatorial Facility was built to store and prepare lunar samples for allocation to scientists..

The present Curatorial Laboratory (Figure 8.5) includes a high-efficiency air-filtration system to remove dust and other particles from the laboratory air, so that it contains less than 1,000 particles (less than 0.5 micron) per cubic foot. A positive air pressure is maintained with respect to the outside, and the floor plan restricts access to vaults, so that the areas which have the most traffic are downstream from the area where the samples are kept. All the building materials that are interior to the building are made of substances that do not shed particles. Cleanliness in the laboratory is important, because the cabinets occasionally have to be opened for cleaning and regloving, and any laboratory dust might carry contaminants into them.

Samples are stored and examined only in nitrogen-filled glove boxes. The nitrogen gas (produced by boil-off of liquid nitrogen) is ultra-pure, with less than 20 ppm Ar, 10 ppm oxygen and 10 ppm water. This relatively inert atmosphere prevents chemical changes in the rocks, such as rust forming on iron grains. Only three materials are allowed to touch the samples: aluminium, Teflon and stainless steel. Teflon overgloves or stainless-steel tongs are used when handling the rocks, so that the rubber gloves do not come in contact with them. Lead from automobile fumes and gold from jewellery are considered two of the worst potential contaminants. The tools used in the cabinets are cleaned with acid and rinsed in Freon to avoid any lead contamination. A Teflon paint, used as a lubricant, was found to be a potential contamination problem and has been removed.

For purposes of avoiding contamination, lunar samples are probably their own best containers so they are not cut up unnecessarily. For storage, the samples are sealed in multiple Teflon bags. Although these bags protect the samples from particulate contamination, they are slowly permeable to gases; consequently, for long-term storage, the bagged samples are also put into stainless-steel sample containers with bolt-on tops. An aluminium gasket between knife edges on the top and the bottom makes a gas-tight seal. These sample containers are stored in nitrogen cabinets in a bank-like vault with extra thick, steel-reinforced concrete walls.

When a sample is needed for study, the sample container is taken from the vault and placed in an airlock entrance to the processing cabinet. The airlock is flushed with nitrogen until the sample can be loaded into the processing cabinet. A protocol for removing various layers of Teflon bags is used to avoid carrying any particles into the processing cabinet. Cabinets are cleaned with liquid Freon to remove all the sample dust between processing. Strict cleaning procedures are also used for all tools and equipment that are used in processing, and this equipment is also subjected to the multiple bag protocol treatment. Only one parent sample is worked on at a time in a cabinet, to avoid cross-contamination of one sample with another. Separate processing cabinets are used for each Apollo mission.

As each rock is processed, it is carefully weighed, photographed, and described. All photographs contain an orientation cube, which relates the orientation of the sub-sample to that of the parent and to the original lunar surface orientation. This is important to the solar flare, cosmic ray and micrometeorite exposure studies. The extensive photographic documentation also tracks the splits of the various lithologies and clasts. Samples are dusted, using a stream of nitrogen gas. Maps are made of the saw cuts and surfaces of the complex breccias by means of binocular microscopes. However, proper petrographic description under these conditions is difficult, and the samples need to be redescribed by later studies outside the cabinet when the sub-samples reach the individual scientists.

Large lunar samples are cut apart with a diamond-edge bandsaw inside a nitrogen cabinet (Figure 8.6). No cooling liquid is used, so the sawed surfaces of some rocks probably get quite hot during this process. A metal smear can be seen on the saw surfaces of some of the hardest rocks. Slabs of the large rocks and the smaller samples are broken up with a stainless-steel chisel. Both sawing and chipping undoubtedly add metallic contamination. Upon completion of a sub-dividing operation, the sub-samples are reassembled into their original positions and photographed as a group for documentation; these photographs are used to construct three-dimensional diagrams and models to help scientists understand the exact position of their sub-sample with respect to those of the sub-samples studied by the other investigators. Sawing done in orthogonal directions is superior to breaking the sample and is preferred, because clasts can be matched up across saw cuts and exact depths of sub-samples can be determined.

Lunar core tubes were first x-rayed and then either cut out of their metal liner with a milling machine or extruded by means of a special device (Figure 8.7). They were then dissected layer by layer. Large particles were documented by photography, and each distinct layer was separated and stored in an individual container. About one-third of the soil was left in the side of the tube, and several peels with sticky methacrylate film were taken of this residual core to remove the layer damaged by dissection and to preserve some particles in their original orientation. Finally, epoxy (diluted with ether) was used to impregnate the remainder, and polished, petrographic thin-sections were made along the entire length of each core.

About 9,000 doubly polished thin-sections have been prepared of lunar samples. After vacuum impregnation with epoxy, the sample is cut with a diamond blade, using alcohol as a coolant, then ground flat with SiC and polished with diamond. The polished sample is epoxyed to a glass slide, cut off, ground to 100 microns thickness and polished by hand to 30 microns, using one micron diamond paste on bond paper. Contamination with terrestrial lead is carefully avoided. Ethyl-alcohol is used instead of water to prevent formation of hydrous phases.

The samples that have been returned from scientists or displays are kept separate from the samples that have never been out of curatorial custody. In case of a fire or hurricane the lunar sample vaults are automatically isolated in 'dead mode', so that no air, smoke or water can penetrate. A portion of the collection (about 50 kilograms) is kept in 'dead' storage at Brooks AFB in San Antonio for extra safety.

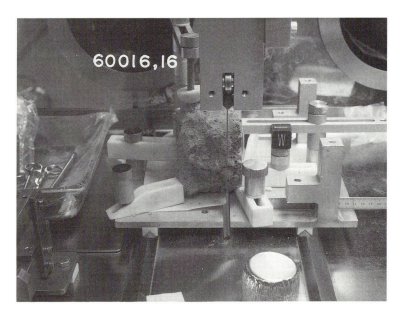

Figure 8.6 Lunar rocks are sawed with a bandsaw in a nitrogen cabinet. This photo shows a typical soil breccia (60016) being cut into slabs to expose new clasts in the interior for scientific study. No lubricant is used in sawing and all fines are saved. The cabinet has to be carefully cleaned after each rock is sawed. The rock is about 6 cm across (NASA S84–40928).

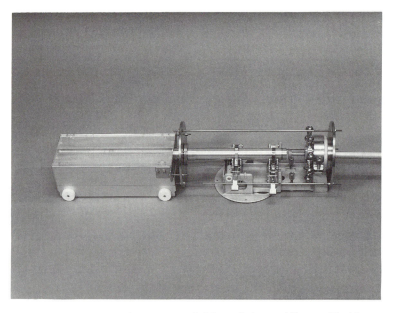

Figure 8.7 Lunar core tubes are extruded from their metal liners with this elaborate device. The receiving cart is 40 cm long and during extrusion has a quartz cover, which is removed for dissection (NASA S80-43518).

Table 8.4 Lunar sample exhibits (circa 1988)

Sites	Samples		Viewers per Year
Adler Planetarium, 1300 South Lake Shore Dr., Chicago, Il 60605	15555,767	74 grams	700,000
Alabama Space & Rocket Museum, Tranquility Park, Huntsville, AL 35807	12065,15	449 grams	600,000
American Museum of Natural History, Central Park West at 75th St. New York, NY 10024	70035,57 60015,179 14305,30	112 grams 135 grams 144 grams	2,500,000
Arizonia Sonora Desert Museum 2021 N. Kinney Rd. Tucson, AZ 85743	75015,54	169 grams	new
Armstrong Air & Space Museum, Ohio Historical Society, 1982 Velma Avenue, Columbus, OH 43211	10017,35	115 grams	70,000
California Academy of Sciences, Golden Gate Park, San Francisco, CA 94118	70035,69	102 grams	1,500,000
Edmonton Space Science Center, Edmonton, Alberta T5M4A1	15555,791	93 grams	300,000
Eisenhower Library, Abilene, KS 67410	15555,461	161 grams	120,000
Hansen Planetarium 15 South State Street, Salt Lake City, UT 84111	15555,464	86 grams	240,000
Houston Museum of Natural Science, Hermann Park, Houston, TX 77030	12018,13	174 grams	1,300,000
International Aerospace Hall of Fame, Balboa Park, San Diego, CA 92101	70035,35	100 grams	250,000
International Space Hall of Fame, Alamogordo, NM 88310	70215,93	113 grams	200,000
Lyndon B. Johnson Library, University of Texas, Austin, TX 78705	15555,161	157 grams	400,000
Kansas Cosmosphere & Discovery Center, 1100 N. Plum, Hutchinson, KS 67501	10020,57 Surveyor scoop	136 grams	350,000
John Kennedy Library, Columbia Point, Boston, MA 02125	15555,50	156 grams	200,000
Michigan Space Center, 2111 Emmons Road, Jackson, MI 49201	15555,54	94 grams	36,000
National Geographic Society Explorer's Hall, Washington, DC 20036	12053,93	134 grams	360,000
Natural History Museum Cromwell Road London, SW7 5BD	60015,87	128 grams	
Noordwijk Space Expo Keplerlaan 3, Noorwijk, The Netherlands	15499,67	177 grams	new
North Carolina Museum of Life and Science, 433 Murray Avenue, Durham, NC 27704	15459,173	8 grams	165,000

continued

Table 8.4 *continued*

Sites	Samples		Viewers per Year
Onizuka Space Centre Kailua-Kona Hawaii	74255,38	117 grams	new
Rieskrater Museum Nordlingen, Germany	66075,26	164 grams	25,000

Space World Inc.
  Yahata-Higashi-Ku
  Kitakyushu-shi, Japan
	12006,1	176 grams	new

Smithsonian Air & Space Museum, Washington, DC 20560	15016,20 60025,53 70051,55 70215, 84 79135,102	65 grams 33 grams 1 gram 38 grams 101 grams	8,000,000

Smithsonian Museum of
  Natural History,
  Washington, DC 20560
	14321,40
	15499,10
	60015,86
	67020,6
	70035,41
	70051,51
	76055,24
		616 grams
		163 grams
		118 grams
		30 grams
		118 grams
		29 grams
		211 grams
			6,000,000

Tycho Brahe Planetarium Kobenhaven, Denmark	75015,53	206 grams	new
United Nations, 799 UN Plaza, New York, NY 10017	14321,86	100 grams	500,000
USAF Museum, Wright–Patterson AFB, OH 45433	67455,1	19 grams	1,500,000
Visitor's Center, Johnson Space Center, Houston, TX 77058	12022,92 76015,143	150 grams 333 grams	800,000
Visitor's Center, Kennedy Space Center, FL 32899	15058,187	130 grams	3,000,000

Visitor's Center,
  Langley Research Center,
  Hampton, VA 23665
	70017,138	160 grams	200,000

Visitor's Center, Lewis Research Center, Cleveland, OH 44135	15015,79	175 grams	140,000
Visitor's Center, Goddard Space Center, Greenbelt, MD 20771	14310,215	100 grams	60,000

Visitor's Center,
  Stennis Space Center,
  NSTL STA, MS 39529
	15015,80	95 grams	50,000

Visitor's Center,
  Wallops Flight Facility,
  Wallops Island, VA 23337
	70035,59	64 grams	60,000

World Intellectual Property Organization,
  34 Chemin Colombettes,
  1211 Geneva 20, Switzerland
	15555,766	105 grams	10,000

8.4 Scientific support

The lunar sample collection is a working collection with about forty active investigative groups. The original set of 2,196 samples has now been sub-divided into more than 84,000 sub-samples. Each sample has a 'data pack' containing complete documentation of the sub-division done in the Lunar Curatorial Facility. All the samples and their weights are tracked by computer, and an annual inventory of every sub-sample is conducted. After scientific study, any remaining sample must be returned to JSC, along with a complete sample history, so that returned samples can be used by other experimenters whenever possible. Samples that are dissolved or destroyed must be carefully documented.

An advisory team of scientists helps the Lunar Sample Curator make recommendations on requests from scientists who wish to study lunar samples. These recommendations are forwarded to NASA

Headquarters for final approval. There has been a free and open world-wide distribution of samples to qualified scientists. Since the samples are most often studied by specialists, a preferred mode of operation for the study of complex samples is by groups of scientists, called consortia, who work together under a leader who coordinates their studies. This coordination helps avoid mixing chemical data, age data and petrographic analysis from the multiple lithologies that are found in most lunar breccias.

Much of the collection has been catalogued more than once in mission and topical catalogues. In 1992 there was still an active effort to update the catalogue, because new lithologies were still being discovered as the breccias were sawed to create new surfaces. Much of the coarse-fine collection, including the particles from the cores, remains to be catalogued.

Some of the important discoveries that have been made during the study of the lunar samples included evidence for an ancient anorthositic crust and an early moon-wide 'magma ocean', a major interval of bombardment about 4,000 million years ago, and a second melting at depth to produce basaltic magma from 4,000 to 3,200 million years ago. It was found that the determination of trace amounts of volatile elements in lunar rocks was an excellent indicator of meteoritic contamination. This technique has since been applied to the study of terrestrial craters and ash layers. On the other hand, the expectation that the history of the particulate radiation from the sun would be recorded at various depths in the lunar core tubes was found to be too difficult in practice, because the regolith is produced by stochastic cratering events instead of by uniform deposition. Scientific studies include trace-element partitioning between mineral phases and melt, regolith formation processes, analysis of volcanic glasses, dating zircons and granite clasts, and modelling Zr/Hf fractionation. The discovery of lunar meteorites has provided the promise of continued new discoveries.

New laboratory techniques that were developed during the study of lunar samples include Sm/Nd and Lu/Hf age dating, U/Pb ion micro-probe age dating, Is/FeO magnetic analysis, Ar 39/40 plateau age dating, and precise, low-level, neutron activation analysis for Au and Ir and rare earth elements. These new techniques have had a significant impact on geochemical studies of many geological collections.

8.5 Educational programmes

NASA sponsors three public programmes that call for the loan of lunar samples for educational purposes. A travelling display programme features samples that range in size from 70 to 160 grams, encapsulated in clear acrylic pyramids. In addition, there are 44 permanent displays set up in museums (Table 8.4), with lunar samples either mounted in nitrogen-filled cases or encapsulated in clear acrylic mounts. Another programme, designed to be used as a science teaching aid in secondary schools, features six small samples in a clear plastic disk, accompanied by written descriptions of each sample, a film, a teacher workbook and other printed material. A third programme, limited to university level petrology classes, uses an educational package of petrographic thin-sections of lunar rocks and soils. These thin-sections are uncovered and require the use of reflecting light microscopes. A detailed descriptive booklet accompanies these educational thin-section sets and provides an excellent introduction to planetary science.

References

1 CADOGAN, P.H., *The Moon–Our Sister Planet*, Cambridge University Press, p. 391 (1981)

2 Proceedings of the Lunar and Planetary Science Conferences 1970-1981, Geochim. et Cosmochim. Acta Supplements 1-16, and 1982, J. Geophys. Res. Supplementary volumes 87-90.

3 FRONDEL, J.W., *Lunar Mineralogy* Wiley-Interscience, New York (1975)

4 SMITH, J.V. and STEELE, I.M., *Lunar Mineralogy*, Amer. Min., **61**, pp. 1059-1116 (1976)

5 BASALTIC VULCANISM STUDY PROJECT, *Basaltic Vulcanism on the Terrestrial Planets*. Lunar and Planetary Institute, Pergamon Press, p. 1286 (1981)

6 MEYER, C., *Petrology, Mineralogy and Chemistry of KREEP Basalt*, Phys. Chem. Earth **10**, pp. 239-260, Pergamon Press (1977)

7 STOFFLER, D., KNOLL, H.D., MARVIN, U.B., SIMONDS, C.H. and WARREN, P.H., *Classification of Lunar Breccias*, Proc. Conf. Lunar Highlands Crust, pp. 51-70. ed. Papike and Merrill, Pergamon Press (1980)

8 HERTOGEN, J., JANSSENS, M.J., TAKAHASHI, H., PALME, H. and ANDERS, E., *Lunar Basins and Craters*, Proc. 8th Lunar Sci. Conf. pp. 17-45 (1977)

9 RYDER, G., *Lunar Anorthosite 60025*. Geochim. et Cosmochim. Acta **46**, pp. 1591-1601 (1982)

10 JAMES, O.B., *Rocks of the Early Lunar Crust*. Proc. 11th Lunar Sci. Conf. pp. 365-393 (1980)

11 DELANO, J.W. and LIVI, K., *Lunar Volcanic Glasses*, Geochim, et Cosmochim. Acta **45**, pp. 2137-49 (1981)

12 MEYER, C., *A Test of the 'Rock Type' Hypothesis*, Proc. 9th Lunar Sci. Conf, pp. 1551-1570 (1978)

13 PAPIKE, J.J., SIMON, S.B. and LAUL, J.C., *The Lunar Regolith*, Rev. Geophys. Space Phys. **20**, pp. 761-826 (1982)

14 MORRIS, R.V., *The Surface Exposure of Lunar Soils*, Proc. 9th Lunar Sci. Conf. pp. 2287-2297 (1978)

15 SCHONFELD, E. and MEYER, C., *The Abundance of Components of Lunar Soils*, Proc. 3rd Lunar Sci. Conf. pp. 1397-1420 (1972)

16 DUKE, M.D. and NAGLE, J.S., *Stratification in the Lunar Regolith*, The Moon **13**, pp. 143-158 (1975)

9

Hazards for the mineral collector, conservator and curator

Frank M. Howie

The purpose of this chapter is to deal with the major risks to safety and health which may arise during the practice of mineralogy, in particular during the collection, processing and conservation of mineral specimens and rocks for institutional collections.[1,2] It is likely that fieldwork[3] produces the greatest risks, and has probably resulted in the largest number of injuries through vehicle accidents and failure to adopt adequate precautions while collecting.

Lack of appreciation of, and over-reaction to, the hazardous effects of chemicals, of which minerals are but a species, appears to be deeply ingrained in the conservation and curation communities in the UK and elsewhere. Besides the inherent toxic and fire hazards associated with many of the chemicals used in processing or preservation, perhaps the most important influences are the human factors. It is often stressed that the use of 'common sense' is the best means of avoiding accidents. In the practice of science, however, lack of training, knowledge or awareness of hazards which may be invisible as well as insidious, ignorance of safety procedures, physiological susceptibility and psychological bias are all factors that may lead to accidents.

In this chapter it is not the intention to cover all the hazardous situations likely to arise out of collecting and processing minerals and rocks. The broad categories of risk are outlined and, with the aid of the reference section, it is hoped provide a basis for further investigation of the safety literature, thereby contributing towards the development of safe-working systems in the field, laboratory, workshop and store.

9.1 Fieldwork: safety considerations

Any type of fieldwork should be considered from the outset as a potentially hazardous undertaking, and steps should be taken to ensure that the collector does nothing which will endanger himself, colleagues, students and, especially, accompanying children. In addition, collectors have a duty of care to members of the public and a responsibility for the property of the owner or occupier of the site.

In the UK there are both common-law considerations and statutory regulations which it is the responsibility of the collector to be aware of and respect. The Health and Safety at Work Act 1974 (UK) and Occupational Safety and Health Act 1970 (USA) put the onus on the employer and employee to take all reasonable, practicable steps to avoid accidents. The Occupiers Liability Act 1984 (UK) specifies that the site occupier should not deliberately put at risk any visitor to his site, whether invited or not.

The references list publications and guides in this area, together with relevant regulations in force at the present time.[4-11] Institutional collectors are bound by instructions their employers have set out, usually in the form of guidance notes or codes of practice, for their employees. Private collectors and the like should be aware of the principles adopted by the institutions, and use these as the basis for their operations in the field.

9.2 Hazards from minerals

Throughout history the processing and mining of ore minerals particularly has led to a wide range of chronic and often fatal afflictions, ranging from skin and lung diseases to cancer.[12] Hippocrates (370 BC) described the symptoms of lead poisoning (lead colic) in Greek ore processors, and Pliny detailed the symptoms and causes of mercurialism and arsenic poisoning in slaves used to mine the ores of these metals. Much earlier, however, in the Ebers Papyrus (1500 BC) and various early Hindu and Chinese tracts, reference was made to the poisonous properties of compounds of lead, copper, antimony, arsenic and mercury. But it was not until the sixteenth century that there was any real awareness of the concept of dose-response to toxic materials.

Agricola dealt with many of the ill effects of poor working conditions in mines in Southern Europe, and highlighted lung diseases caused by ore dust and ulceration caused by exposure to arsenic minerals.[13] Ramazzini described and characterized several dozen occupational health diseases, many of which were caused by exposure to minerals and rocks. Many of the descriptive terms then used have passed into common parlance, e.g. 'masons' disease', 'potters' rot', 'rock tuberculosis', 'stone cutters' asthma' (later becoming stannosis), beryliosis, siderosis.[14]

As late as the 1920s in California miners suffered chronic and acute symptoms after short exposure to the insidious mixture of metallic mercury and dust found in cinnabar mines, and today of course we have only just begun to realize the extent of crippling illnesses and premature mortality caused by both occupational and non-occupational exposure to asbestiform minerals. Uranium mining in the USA now requires controlled time exposure of miners enclosed in lead-shielded diggers and tractors, thus exemplifying the current concern for personal safety.

With such a notorious history, one might expect minerals in collections to present grave problems. This is, however, not at all the case, and aside from deliberate or accidental ingestion of some of the more highly toxic minerals, there are hardly any reports of occupational illness caused by minerals in collections to date. Such collections are not without risk, and in this section the main factors are considered – first from radioactive minerals and second from toxic minerals.

9.2.1 Radioactive minerals

Certain types of work with apparatus used to generate ionizing radiation and sources of radiation require registration under UK and US Health and Safety legislation. Special provisions for dealing with the public exhibition of radioactive substances have been passed in the UK.[15] However, limits of exposure, i.e. dose limits for members of the public, are at present under review. The International Commission on Radiological Protection (ICRP) believes at present that the limit should be 1m Sv per year, but it is suggested that a subsidiary dose limit of 5m Sv per year for some years is allowable, provided the average annual effective dose limit over a lifetime does not exceed 1mSv per year.

9.2.2 Storage of radioactive minerals

Brunton *et al.*[16] and Dixon[17] give detailed descriptions of some of the main hazards likely to be encountered while storing, curating and exhibiting radioactive minerals. Exposure to levels above 7.5 µSv per hour, the present limit for employees in non-designated areas, is not likely to occur under normal circumstances; but with large specimens and large collections, e.g. dealers' stocks, it may be necessary to monitor local radiation levels. As a rule, exposures to radiation levels should be as far below this figure as possible. Table 9.1 lists some of the radioactive minerals likely to be met with in collections, together with activity data where known or relevant to safe storage. The dangers of storing radioactive minerals together in one area may outweigh the curatorial advantages. Firstly, radiation levels near or even above the dose limit may be present in the vicinity and, secondly, there has been concern about the build-up of highly toxic radon and its stable daughter isotopes in radioactive mineral stores to levels in excess of safety limits. It is generally recommended that enclosed stores should be extract-ventilated to remove radon.

As a matter of course monitoring of radiation levels (alpha, beta and gamma radiation) should be carried out. In addition, the activity due to radon decay should be measured.

The criteria[18] for control and, possibly, designation of the store or an area in it as a controlled or supervised area are dependent on three factors:

(a) The total activity being handled.
(b) The activity in air in the area.
(c) The surface contamination level in the store.

Expert advice on the above aspects should be obtained from an appointed radiological protection adviser or safety officer with specialist training in radiological safety.

9.2.3 Handling radioactive minerals

Brunton *et al.*[16] and Dixon[17] suggest that exposure to radiation during the routine handling of uraninite and other highly radioactive minerals is negligible in comparison with the notified dose limits. Uranium minerals are toxic not so much on account of their radioactivity but because of the physiological effect of soluble uranium compounds on the kidneys.[19]

Prolonged exposure to uraninite and several other uranium minerals is hazardous, and it is likely that guidelines based on those governing the handling of unsealed radioactive substances will apply to handling many radioactive minerals in collections before too long. The basic rules of personal hygiene during the handling of radioactive minerals should include:

● Use of protective clothing, i.e. gloves (PVC, latex or polythene, disposable type) plus toxic dust/high efficiency filter respirator (to BS4555) for handling active species.

Table 9.1 Some radioactive minerals

Name	% Radioactive element	Specific activity range Bq g⁻¹
Andersonite	up to 40% U	
Arsenuranylite	up to 60% U	
Autunite	up to 50% U	3 – 51
Bayleyite	up to 40% U	
Becquerelite	up to 30% U	+ 5,000
Betafite	up to 30% U	
Billietite	50–70% U	
Brannerite	27–43% U	
Carnotite	53–55% U	68–1,040
Clarkeite	up to 55% U	
Coffinite	up to 75% U	
Curienite	up to 45% U approx.	
Curite	up to 50% U	
Davidite	10–15% U	
Euxenite	1–12% U	310–720
Fourmarierite	up to 50% U	
Fergusonite	1–7% U	
Gummite	Th + U up to 50%	
Heinrichite	40% U approx.	
Hugelite		
Huttonite	up to 70% U	
Ianthinite	up to 60% U	
Johannite	up to 50% U	
Kasolite	50% U approx.	
Liebigite	50% U approx.	
Metaautinite	up to 50% U	
Metatobernite	up to 50% U	
Metauranopilite	up to 80% U	
Microlite	up to 30% Th	
Moluranite		
Monasite	up to 20% Th	
Natroautinite	up to 50% U	
Pitchblende	45–90% U	3,000–8,000
Polycrase	up to 10% U	
Pyrochlore	up to 7.5%	
Rabbitite	30% U approx.	
Rutherfordine	75% U approx.	
Samarskite	8–17% U	
Schoepite	up to 60%	
Sharpite	up to 70% U	
Soddyite	up to 60% U	
Swartzite	up to 40% U	
Thorianite	up to 45% U + Th	
Thorite	u to 10% U + Th	
Tobernite	47% U	
Tyuyamunite	45% U approx.	
Uraninite	45–90% U	3,200–8,500
Uranocircite	up to 50% U	
Uranopilite	up to 80% U	
Uranospathite	up to 50% U	
Uranosphaerite	30% U approx.	
Uranotile (uranophane)	50% U approx.	
Vanuranilite	60% U approx.	
Vanuralite	45% U approx.	
Voglite	30% U approx.	
Wolsendorfite	30%–40% U approx.	
Yttrialite	5–6% Th	
Zeunerite	48–50% U	
Zippeite	70% U approx.	
Zircon	up to 3% U	

- Obligation to wash hands after handling all other radioactive mineral specimens.
- Monitoring hands after washing and after handling dusty or friable specimens.
- Not handling radioactive minerals when hands are cut or skin broken.
- Not eating, drinking or smoking in areas where radioactive minerals are stored or handled.

9.2.4 Toxic minerals

Puffer[19] lists 200 toxic minerals, Lof[20] gives about 70 in his world minerals chart and Brunton *et al.*[16] outline the toxic properties of about 40 minerals. Each authority lists many minerals the others do not include, although Lof's and Brunton's lists are not designed to be exhaustive.

Table 9.2, culled from these sources, together with data from Hunter,[12] Waldron[21] and others, contains those minerals considered to be moderately to highly toxic by ingestion, inhalation (of dust) or prolonged or repeated skin contact. There are undoubtedly many other toxic minerals, and some on this list are so rare, or occur in such minute quantities, that they really do not constitute any significant degree of risk to health.

Accidental acute poisoning by ingesting toxic minerals is unlikely, except for lead, arsenic, mercury and thallium compounds. The lethal dose of arsenic (as As_2O_3) may be as low as 20 mg. Chronic effects may also be a definite risk for curators and others who handle arsenic or oxidized-lead minerals regularly. The UK Health and Safety Executive considers that arsenic and its inorganic compounds should be regarded as human carcinogens. In addition, it is proposed that anyone exposed to arsenic compounds other than very occasionally should be subject to medical surveillance. Ingestion of as little as 20 mgs of arsenic has been known to be lethal to a child. The addition of acid to arsenic minerals should be avoided, because of the risk of arsine liberation; exposure to concentrations as low as 10 ppm has proved fatal.

Thallium minerals should all be regarded as highly toxic, as they are likely to be rapidly assimilated by the body. The lethal dose for thallium as the oxide is 100–200 mg for an adult. Note that Clericis solution, a molar mixture of thallium malonate and formate used for mineral flotation analysis, is probably even more toxic, because of its ability to penetrate the skin rapidly. Arsenic and thallium minerals should be handled with suitable gloves, i.e. latex or neoprene. All mechanical processing of minerals containing toxic elements, including cutting and grinding of most of the minerals listed in Table 8.2, should be carried out in an effective fume cupboard.

Table 9.2 Moderate to highly toxic minerals

Mineral	Toxic element(s)	Ingestion	Inhalation	Skin
Adamite	As	*	*	*
Anglesite	Pb	*	*	O
Annabergite	Ni,As	*	*	*
Antimony (native)	Sb	*	*	O
Arsenic (native)	As	*	*	**
Arsenopyrite	As	*	O	*
Arsenolite	As	*	*	**
Asbestos minerals	-	*	*	O
Avicennite	Te	*	*	*
Bequerelite	U	×	*	×
Behoite	Be	*	*	×
Bieberite	Co	*	×	**
Boleite	Pb	*	*	O
Boracite	B	*	×	O
Borax (+ all sol borates)	B	*	×	O
Bournite	Pb Sb	*	*	O
Bromellite	Be	*	×	**
Bunsenite	Ni	*	*	**
Calomel	Hg	*	*	**
Carnotite	U, V	×	*	**
Carrobite	F	*	×	O
Carlinite	T1	*	*	*
Cerrantite	Sb	*	*	*
Cerrusite	Pb	*	*	O
Chromite	Cr	×	*	**
Cinnabar	Hg	*	*	O
Claudetite	As	*	*	*
Conichalcite	As	*	*	*
Crocoite	Pb	*	×	O
Crooksite	Tl	*	*	*
Curite	Pb, U	*	*	*
Cristobalite	Si	O	*	O
Descloizite	Pb, V	×	*	**
Emerald	Be	O	*	O
Erythrite	As	*	*	O
Eskolaite	Cr	*	*	*
Finnemanite	Pb, As	*	*	*
Fieldlerite	Pb	*	*	O
Frankdicksonite	Ba, F	*	×	O
Greenockite	Cd	*	*	poss
Goslarite	Zn	*	×	O
Georgiadesite	Pb, As	*	*	*
Hawleyite	Cd	*	*	poss
Heliophyllite	Pb, As	*	*	*
Jamesonite	Pb, Sb	*	*	O
Karelianite	V	×	*	**
Kermesite	Sb	*	×	O
Kasolite	Pb, U	*	*	O
Koettigite	Zn As	*	*	**
Lead (native)	Pb	*	*	*
Legrandite	Zn, As	*	*	*
Linarite	Pb	*	*	O
Loellingite	As	*	*	×
Lorandite	Tl	*	*	*
Malachite	Cu	O	*	O
Melanterite	Fe	*	×	O
Mimetite	Pb, As	*	*	*
Monteponite	Cd	*	*	*
Montroydite	Hg	*	*	**
Mercury (native)	Hg	*	*	**
Niccolite	As, Ni	*	*	×

Table 9.2 *continued*

Mineral	Toxic element(s)	Ingestion	Inhalation	Skin
Nickel (native)	Ni	*	×	**
Nitrobarite	Ba Ni	*	*	O
Orpiment	As	*	*	**
Pharmacolite	As	*	*	**
Phoenicochroite	Pb, Cr	*	*	**
Phosgenite	Pb	*	*	?
Proustite	As	*	*	×
Pierrotite	T1	*	*	*
Pyragyrite	Sb	×	*	×
Pyromorphite	Pb	×	*	×
Phenakite	Be	×	*	×
Penfieldite	Pb	*	*	×
Paratellurite	Te	*	*	×
Quartz	Si	O	*	O
Realgar	As	*	*	**
Retgersite	Ni	*	*	**
Routherite	T1	*	*	*
Senarmontite	Sb	*	*	×
Schultenite	Pb, As	*	*	**
Spherocobaltite	Co	×	×	**
Siderite	Fe	O	*	O
Skulterudite	Co, As	*	*	**
Stibnite		×	*	O
Shcherbinaite	V	×	*	**
Selenium (native)	Se	*	*	**
Selenolite	Se	*	*	**
Tellunite	Te	*	×	×
Tennanite	As	×	*	O
Tetrahedrite	Sb	×	*	O
Thorianite	U	*	*	×
Tobernite	U	*	*	×
Uranite	U	*	*	×
Vanadite	Pb, V	×	*	poss
Vrbaite	Tl	*	*	*
Wulfenite	Pb	*	×	O

O Not known to be toxic via this route.
× Unlikely to be toxic via this route.
* Likely to be highly toxic by this route.
** Known skin sensitizers.
Minerals in italics could present problems during handling.

9.2.5 Hazards of mineral processing

Several mineral dusts will cause lung disease where operators are regularly exposed to high levels; Waldron[21] specifies siderosis (from processing iron ore), stannosis and asbestosis. The likelihood of the mineral preparator or conservator developing any of these conditions is exceedingly remote so long as effective means are used to control dust levels during rock-cutting and mechanical preparation. The use of approved respiratory protection (to HSE approved standard in the UK or NIOSH approved in the USA) for toxic particulates is the minimum requirement for occasional exposure. Puffer[19] lists some of the minerals likely to give rise to dust hazards when processed, but few present any respiratory hazard to the casual handler. Where mechanical preparation, rock-cutting and trimming, are routine it is essential to use some form of effective local exhaust ventilation (LEV) as many rock types contain free silica, iron oxide, asbestos and other toxic components. Without an LEV system, dust levels produced during this type of work will rapidly attain levels far in excess of the maximum exposure limits and occupational exposure standards.

Current research into many of the lung diseases caused by occupational exposure to mineral dusts demonstrates that smokers exposed, for example, to silica or asbestos dust have considerably increased mortality rates when compared to non-smokers exposed to similar risks. Rock types and mineral association containing free silica are particularly hazardous. Sandstones, slates, shales, soapstone, some limestones and many igneous rocks, onyx, jaspar, agate and turquoise come into this category. Talc, soapstone, green stone, steatite and serpentine often contain appreciable quantities of asbestiform minerals.

9.3 Laboratory safety and chemical handling

Surveys of accidents in medical and university laboratories[22,23] show that the majority (over 60%) were caused by cuts and abrasions from glassware, knives, scalpels, etc. Chemical accidents, poisoning, gassing and explosions account for less than 5% of accidents, and the balance of 35% were caused by slips, falls, faulty handling and lifting methods.

9.3.1 Fire and explosion hazards

Situations to avoid when handling highly flammable organic solvents include the use of electric stirrers (non-explosion-proofed) for the preparation of consolidants; the use of consolidants near microscope lamp transformers, including fibre-optic light sources; and the use of a hot plate, sand bath or oil bath to heat flammable organic compounds. Avoid storage of organic solvents in domestic refrigerators; the build-up of vapour in the refrigerator can be ignited by the thermostat or interior lamp contacts. Fatalities and serious injuries have been caused by refrigerator doors blowing out and hitting the person opening the door.

Solvent-soaked tissues and rags in open bins are a common cause of fire. *Always use* closable metal bins (several proprietary types are available but expensive); a small all-metal bin is adequate. Only use small amounts of solvents in the open laboratory; the less on the bench the less likely for a serious fire to develop. *Never* smoke or allow visitors to smoke in laboratories.

9.3.2 Health hazards

The principal routes for chemicals to enter the body are via the eye, mouth, respiratory tract, through lacerations and by skin adsorption. Large concentrations of solid, liquid or vapour phases of several chemicals will cause severe acute symptoms, ranging from nausea, headache, vomiting (these are mainly flight reactions, i.e. warnings for the body to pull out and avoid the situation) to broncho-spasm, neurotoxic effects, coma (generally when it is too late to take avoiding action or where the chemical is very fast acting) and death. Of great concern to geologists have been recent reports of deaths caused by acute exposure to high concentrations of hydrofluoric acid vapour.[24]

Avoidance of acute injury by solvents or corrosive chemicals is simple enough: enclose the process in a fume cupboard or hood, use appropriate protective screen and use goggles, face mask and gloves when handling acids or alkalis. Always carry out operations with moderate or high risk chemicals in fume cupboards or under effective local exhaust ventilation systems. Clark *et al.*[25] and Howie[26] have reviewed the use of ventilation in conservation laboratories and workshops. It is perhaps most important to ensure that new staff and volunteers are given extensive training in chemical handling techniques. Do not assume that technically qualified staff have received adequate grounding in laboratory safety. Last but not least, do not allow food or drink to be consumed where chemicals are handled.

9.3.3 Incompatible chemicals

Many types of chemical reactions do not require any heat, light or catalyst to initiate them; simply mixing certain liquids, gases or even solids may cause fire, explosion or serious injury. Accidental mixing of incompatible materials can be avoided so long as one is aware of the likely consequences. Conservators and preparators often have their favourite 'cocktail' of solvents or acids, etc., for cleaning specimens and preparing consolidants. During preparation, sequential treatment of material with, for example, acids and peroxides can be hazardous, especially where specimens are not adequately washed between treatments. See Bretherick[27] for details of many of the known chemical incompatibilities.

9.4 Controlled exposure to hazardous substances

Anyone using chemicals for any purpose, or processing minerals, should be aware of the requirements introduced by the Control of Substances Hazardous to Health Regulations (1988).[28] A great many chemicals were, up to a few years ago, used without any real control of exposure, for example, asbestos, vinyl chloride and lead compounds. Some, such as formaldehyde, isocyanates and styrene, have recently been shown to be potentially very harmful over the short and long term at levels of exposure just above occupational exposure limits.

9.5 Physical hazards

Conservators, curators and preparators are likely to encounter a range of physical hazards, and five main categories will be considered here:

- Noise.
- Vibration.
- Non-ionizing radiation.
- IR radiation.
- Handling and lifting.

9.5.1 Noise

Impulse noise is the most hazardous type, and exposure to repeated gunfire and drop forging without adequate hearing protection can quickly lead to loss of hearing. In these circumstances sound pressure levels of 150 dB(A) or greater will occur. Hammering hard igneous rocks may well produce levels of 110–130 dB(A) and mechanical rock-cutting, splitting and grinding equipment may produce noise levels in excess of 100 dB(A).

The present minimum action level for exposure to noise at work is 85 dB(A) on a sustained basis, that is averaged out over an eight hour day.[29] The dB(A) notation is a logarithmic scale, and for every three dBA rise the noise level or sound pressure will be effectively doubled.

9.5.2 Vibration

Many of the hand-held tools used for the preparation of rocks and fossils produce high levels of vibration.[30] Two types of injury can occur with prolonged use of such equipment. The first is the harmless production of cysts in the bones of the wrists of those using pneumatic tools, drills and chain saws, etc. The second, vibration-induced white finger (VWF), is more serious, and in extreme cases can lead to partial disablement. Instances of susceptibility to VWF in preparators and conservators have recently been discovered. The greatest hazard appears to occur where electro-vibratory and pneumatic tools are used without sufficient damping of the object being worked on or with inadequate damping of the handpiece itself.

9.5.3 Non-ionizing radiation

Included in this category are ultraviolet (UV), infrared (IR) and radio-frequency radiation. UV and IR generating equipment is used frequently in laboratory work.

UV radiation (between 200–400nm) is divided into:

UV-A (near UV or 'black' light)   400–315 nm
UV-B (erythemal region)   315–280 nm
UV-C (far UV)   280–100 nm

UV below 250 nm will dissociate oxygen in air to ozone, and below 160 nm will cause nitrogen to react with oxygen forming nitrogen oxides.

Lamps, including mercury, metal halide and inert gas as well as flash and fluorescent tubes, emit UV in varying wavelengths. High-level UV illumination lamps may be fitted with double envelopes, the outer envelope being the UV filter. This outer case can be broken without it being noticed, and eye injuries have occurred through exposure to UV-B

and UV-A. No exposure to UV light is entirely safe; shorter wavelength UV affects the epidermis, whereas at wavelengths above 300 nm deeper effects on the dermis are possible. Wavelengths of 300 nm will affect the lens of the eye.

Acute effects of exposure (erythema or skin-reddening and 'arc-eye') usually occur several hours after exposure. Chronic effects include loss of skin elasticity and cancer, and, in the eye, cataracts and lens opacity. The limits for exposure to UV applying to both skin and eye are based on time-weighted irradiation levels at 270 nm. These range from 8 hours per day at 0.1 W cm^{-2} to 0.1 second per day at 30,000 W cm^{-2}. For UV-A the level is set at 1 m W cm^{-2} per day for duration of exposure of more than 16 minutes.

9.5.4 IR radiation

From 700 nm to microwave, radiation at 300 G N$_3$ has the most marked effect on the eye; irradiation at 2,000 nm causes a temperature rise of $45\,^{\circ}$C in the cornea in 2–3 milliseconds, and this causes intense pain. Cataracts may be an effect of repeated exposure to IR. the high-intensity IR lamps used for drying purposes generally operate at wavelengths of <700 nm to 1,500 nm and will not normally have any harmful effects. There are as yet no standards for exposure to IR sources (except IR lasers).

9.5.5 Handling and lifting

The manipulation of geological specimens presents special problems, as they can, at one time, be heavy, fragile and irregularly shaped. It is probable that the majority of museum geologists have suffered acute or chronic muscular or skeletal strains through handling specimens, whether in the field or the collection. The approach to lifting large geological specimens requires three basic considerations:

- Conversion of an often irregular object into one which is more easily grasped and manipulated.
- Adoption of strain-free lifting techniques.
- Use of manual handling equipment.

References

1   BUIST, A.S., 'Human health hazards from volcanic eruptions: approach to evaluation of a new environmental health hazard', *Health*, 76, pp. 1–90 (1986)

2   MCCANN, M., *Artists Beware* Watson-Gupthill, New York (1979)

3   THOMASON, J.G., *Guidance Note: Safety in Fieldwork* National Environmental Research Council, Swindon (1983)

4   NICHOLS, D., *Safety in Biological Fieldwork* Institute of Biology, London (1984)

5 DEPARTMENT OF EDUCATION AND SCIENCE, *Safety in Outdoor Pursuits* HMSO, London (1977)

6 BRITISH MOUNTAINEERING COUNCIL, *Safety on Mountains* BMC, London (1976)

7 *The Law Relating to Safety and Health in Mines and Quarries* Parts 1-4 and supplements (1972-1979) HMSO, London

8 LANEY, J.C., *Site Safety*, Construction Press, London (1982)

9 HORNER, P.C., *Earth Works* Thomas Telford, London (1978)

10 BRITISH STANDARDS INSTITUTION. *BS6408: 197 specification for clothing made from coated fabrics for protection against wet weather* BSI, Milton Keynes (1983)

11 TURNER, A.C., *The Traveller's Health Code* Lascelles, London (1979)

12 HUNTER, D., *Diseases of Occupations* Hodder and Stoughton, London (1979)

13 AGRICOLA, *De Re Metallica*, Basiline (1556)

14 RAMAZZINI, *De Morbis Artificum Diatriba*, Modena (1700)

15 NRPB, *Code of Practice for the display of Sources of Ionizing Radiation at Exhibitions*, National Radiological Protection Board, Didcot (1966)

16 BRUNTON, C., BESTERMAN, T.P. and COOPER, J., *Guidelines for Curation of Geological Collections*, Misc Paper No. 17, Geological Society, London (1985)

17 DIXON, D., *Radiation Hazards to Collectors of Geological Specimens Containing Natural Radioactivity NRPB-R131*, National Radiological Protection Board, Didcot (1983)

18 HEALTH AND SAFETY COMMISSION, *The Protection of Persons against Ionizing Radiation arising from any Work Activity: The Ionizing Radiation Regulations 1985*, HMSO, London (1985)

19 PUFFER, J., 'Toxic Minerals', *Mineralogical Record*, 11, pp. 5-11 (1980)

20 LOF, P., *Minerals of the World*, Elsevier, Amsterdam and New York (1985)

21 WALDRON, A., *Lecture Notes on Occupational Medicine*, Blackwell Scientific, Oxford (1985)

22 DEWHURST, F., 'Accidents, Safety and First Aid training in Laboratories', *International Environmental Safety*, 1, pp. 11-13 (1983)

23 HERMAN, B.A., 'Morbidity and Mortality Studies', in *Special Publication No. 52 Health and Safety in the Chemical Laboratory-Where do we go from here?* Royal Society of Chemistry, London, pp. 15-29 (1984)

24 HOWIE, F., 'Safety considerations for the Geological Conservator in Conservation of Geological Material', *The Geological Curator*, 4, pp. 379-403 (1987)

25 CLARK, N., CUTTER, T. and MACCRANE, J., *Ventilation: a Practical Guide*, Centre for Occupational Hazards, New York (1984)

26 HOWIE, F., 'Exhaust ventilation systems for museums: some basic principles', *Conservation News*, 29, pp. 17-19 (1986)

27 BRETHERICK, L., *Handbook of reactive chemical hazards*, Butterworths, London (1986)

28 HEALTH AND SAFETY EXECUTIVE, *COSHH Approved Code of Practice*, HMSO, London (1989)

29 HEALTH AND SAFETY EXECUTIVE, *Noise at Work: Guidance on regulations Pts 1 & 2*, HMSO, London (1989)

30 CONSTABLE, P., 'Occupational Health in Museums', in Howie, F.M.P. (Ed.), *Safety in Museums and Galleries*, Butterworths, London, pp. 67-74 (1987)

Appendix I

Effects of construction materials on rock and mineral collections

Table AI.1 Effects of construction materials on rock and mineral collections (buildings, storage and exhibition).

Construction materials	Pollutants and volatiles	Likely effects	Effects likely on	Action	References
Concrete (new)	Alkaline aerosols	Corrosive environment	Metals, sulphides, silicates, etc.	Delay occupation of newly constructed facilities for 3–6 mths	1
	Alkaline dusts	Sticky dusts	All minerals		2
Plaster (new)	Moisture, dusts	High RH	Metals, sulphides hydrates, etc.	Seal plaster as soon as building permits; thoroughly clean before occupation	
Cavity fillers	Aldehydes, CFC, organic acids, ammonia	Corrosive environment	Metals, oxides, sulphides, etc., carbonates	Avoid corrosive types; specify inert gas-expanded fillers	
Forest products and uncoated timbers	Volatile organic compounds are slowly and continuously emitted at very low rates by all timbers; all will yield organic acids (usually acetic, some formic) under uncontrolled conditions (high T and high RH). Some are active at normal temps	Most timbers likely to have corrosive effect in uncontrolled environment over short periods. Little evidence of long-term effects with most timbers (except those below) where environment controlled to habitable levels (i.e. 15°–25°C and 20–70 (percent; RH)	Metals, oxides/sulphides, etc.	Ensure that any timbers used are correctly dried to moisture content for climate zone. Kiln/oven dried, steam-treated and pressure-impregnated timbers are more likely to become corrosive than naturally dried timbers; where timber used, avoid exposure of end-grain and maintain ambient T at 20°C. Use softwoods wherever possible, as lower hemicellulose content and lower rate of hydrolysis to acetic acid	3 4 5
Basswood	Organic acids	Corrosion	Metals, oxides, sulphides, etc.,	Avoid use	6
Birch	Acetic (?) acid	Corrosion	Some calcereous minerals, some carbonates	Avoid use	

Material	Hazard	Effect	Affected	Recommendation	Ref
Chestnut (sweet)	Organic acids	Corrosion	Metals, oxides sulphides, etc.	Avoid use	5
Oak (all types)	Acetic and formic acids, yield accelerated at temps >25°C	Corrosion	Some calcerous minerals, some carbonates	Avoid use. Coatings will not prevent emission of acidic vapours.	5, 7
Cardboard	Organic acids, formaldehyde	Corrosion	Metals, oxides sulphides, etc.	Always use acid-free cardboard	8
Chipboard (Sundaela and Particle boards)	Aldehydes, acid catalysts, organic acids, wood volatiles, phenol and ammonia	Corrosion	All minerals	Avoid use	9
Fibre-board (MDF)	Acids (volatile)	Corrosion by contact	Metals, oxides, sulphides, etc.	Avoid direct contact. Use only those bonded by starch	10
Hardboard	Acids, formaldehyde and phenol	Corrosion by contact	All minerals	Avoid use	11
Paper (acid-processed)	Organic and inorganic acids, sulphur dioxide	Corrosion	All minerals	Always use acid-free paper	
Plywoods	Organic acids, acid catalysts, phenol, ammonia and formaldehyde	Corrosion	All minerals	Avoid wherever possible, especially birch-ply and oak-faced boards	
Cork (natural and composite)	Aldehydes, organic acids, sulphur compds	Corrosion, sulphidization	Metals, oxides, sulphides, etc.	Avoid use generally	11
Plastics					
Acrylic plastics	None likely				
Polyethylene	None likely, some may emit volatiles			Specify non-chemically polymerized types	4, 12
Polypropylene	None likely				4
Polyurethane	None likely				4
Polystyrene	None likely				
Polyester	If under-cured, emits formic acid	Corrosion	Calcareous minerals	Ensure fully-cured before use	4

Continued

Table AL.1 *continued*

Construction materials	Pollutants and volatiles	Likely effects	Effects likely on	Action	References
Nylon	Nylon 6 uses acetic acid as MW regulator	Corrosion	Calcareous minerals	Avoid use	13
Formaldehyde, phenol- and urea-	If under-cured emits formaldehyde, ammonia, phenol, acid catalysts	Corrosion	All minerals	Avoid use wherever possible	4, 8
ABS, moulded	None likely				
Epoxy resins	Volatile curing agents			Ensure fully cured before use.	
Polyvinyl chloride	Likely to decompose when exposed to heat and UV (e.g. sunlight) and emit hydrogen chloride.	Corrosion, especially in close confinement	All minerals	Avoid use wherever possible	
Paints, adhesives and lacquers, etc.					
Acrylics: emulsions,	Methacrylic/acrylic acid, solvents,	Corrosion	Metals	Do not use to coat sulphides, etc., and metals	9
solvent based paints	Acetic acid	Corrosion	Metals, oxides some calcereous minerals	Avoid use	14
Animal glues	May emit volatiles when old/hydrolysed	Corrosion	Metals	Do not expose to high T and high RH	1
Alkyd paints (stoving paints)	Aldehydes, organic ids	Corrosion	Metals, oxides sulphides, etc.	Avoid use	4
Epoxy paints	Solvent fumes	None likely		Allow several weeks for drying	11
Epoxy resins	Volatile curing agents			Avoid use as adhesive on delicate minerals	
Latex paint and coatings	Ammonia, hydrogen sulphide (when aged)	Corrosion	Metals, some sulphides, etc.	Avoid use	10, 11
Natural resins	Organic volatiles	Mostly non-corrosive			
Shellac	None likely				

Material	Volatiles/source	Process	Affected materials	Recommendation	Ref.
Starch	None likely				
Oil-based paints and varnishes	Organic acids, aldehydes	Corrosion	All minerals	Avoid use	4, 11
Urethane-based paints and coatings	Solvent fumes	None likely		Allow several weeks for drying	11
Cellulose acetate	Acetic acid at high T	Corrosion	Metals	Avoid use	9
Cellulose nitrate	Spontaneous decomposition	Corrosion	Metals	Avoid use	
Silicone-based sealants	Acetic acid during curing process	Corrosion	Metals	Allow several days for curing	15
Rubbers, vulcanized and synthetic	Volatile sulphur compounds. None likely if fully cured	Sulphidization	Some metals, sulphides, etc.	Avoid use	4
Polyvinyl acetate, emulsions, solvent-based	Acetic acid from unstabilized PVA	Corrosion	Metals, oxides, carbonates, etc. Some calcareous minerals	Avoid use generally as adhesive or consolidant. Use only PVA with inhibitors and stabilizers	9, 10
Polyvinyl butyral	None likely				
Cyanoacrylates	Volatiles on curing	Staining and bloom	Iron-rich rocks	Avoid use as adhesive for minerals.	
Textiles and fabrics	Pre-treatment acids, pest-proofing and flame-retarding agents			Use only fabrics and textiles that are rated for use for object conservation	16
Cotton	Organic acids	Corrosion	Metals	Avoid use	9
Felts	Hydrogen sulphide	Sulphidization	Sulphides, etc., metals, oxides	Avoid use wih sulphides, metals and oxides	16
Hempcloth	Organic acids	Corrosion	Metals	Avoid use	10
Leather	Hydrogen sulphide organic volatiles	Sulphidization	Sulphides, etc., metals, oxides	Avoid use	9
Silk	Processing agents	Corrosion	Metals	Avoid use	10
Wool	Hydrogen sulphide carbonylsulphide	Sulphidization	Sulphides, etc., metals	Avoid use	9, 10

References to Appendix I

1 HOWIE, F.M.P., 'Museum climatology and the conservation of palaeontological material', in *Special Papers in Palaeontology*, No. 22, pp. 103-25 (1979)

2 TOISHI, K. and KENJO, T., 'Some aspects of the conservation of works of art in buildings of new concrete', *Studies in Conservation*, 20, pp. 118-22 (1975)

3 HAYGREEN, J.G. and BOWYER, J.L., *Forest Products and Wood Science*, Iowa State University (1982)

4 DONOVAN, P.D. and STRINGER, J., 'Corrosion of metals and their protection in atmospheres containing organic acid vapours', *British Corrosion Journal*, 6, pp. 132-8 (1971)

5 FARMER, R.H., 'Corrosion of metal in association with wood. Part 1, corrosion by acid vapours from wood', *Wood August*, pp. 326-8 (1962)

6 GRAHAM, R.D., WILSON, N.M. and OTENG-AMOAKS, A., 'Wood-metal corrosion: An annotated survey', *Oregan State University Forest Research Laboratory, Research Bulletin,* 21, Corrallis, OSU (1976)

7 PACKMAN, D.F., 'The Acidity of Wood', *Holsforschung*, 14, pp. 178-83 (1960)

8 RANCE, V.F. and COLE, H.G., *Corrosion of metals by vapours from organic materials, a survey*, HMSO, London (1958)

9 BLACKSHAW, S.M. and DANIELS, V.D., 'Selecting safe materials for use in the display and storage of antiquities', *Preprints 5th Triannual Meeting ICOM Committee for Conservation 78/23/2*, pp. 1-9 (1978)

10 HNATIUK, K., 'Effects of display materials on metal artifacts', *Gazette of the Canadian Museums Association Summer-Fall*, pp. 42-50 (1981)

11 MILES, C.E., 'Wood coatings for display and storage cases', *Studies in Conservation*, 31, pp. 114-24 (1986)

12 WALLER, R., personal communication

13 DONOVAN, P.D., 'Corrosion of metals by plastics', *Corrosion*, 2 (1976)

14 DONOVAN, P.D. and MOYNEHAN, T.M., 'The corrosion of metals by vapours from air-drying paints', *Corrosion Science*, 5, pp. 803-24 (1965)

15 DONOVAN, P.D. and STRINGER, J., 'Corrosion of metals and their protection in atmospheres containing organic acid vapours', *British Corrosion Journal*, 6, pp. 132-98 (1971)

16 ODDY, W.A., 'The corrosion of metals on display', in *Conservation in Archaeology and The Applied Arts* Stockholm, pp. 235-7 (1975)

Appendix II

Bob King

Collecting rocks and minerals

The extraction of good crystallized minerals can be difficult, especially where, because of quarrying and mining operations, rocks are damaged by vibration and shock. However, crystals which have grown completely enclosed in, say, clay, are simple to extract safely.

Cavities lined with crystals (vugs) are a good source of well-developed minerals. Special care should be exercised in the extraction of minerals showing a well-marked cleavage, as rough handling may break them or cause the development of incipient cleavages.

Destructive collecting from natural exposures is now actively discouraged, and the rapidly advancing branch of mineral collecting known as micromount mineralogy, which is confined principally to old mine dumps, is taking its place. The standard method of collecting micromounts is to take samples of promising looking material from a recognized site. Once in the laboratory, the material is split and examined under the microscope, and selected micromounts are split off the sample and mounted for storage in small plastic boxes.

So far as hard rock samples are concerned, the collector must decide whether unweathered material is required, for example, for microslide preparation, or whether textural details are required. If the latter, a good photograph should suffice.

Locality data are of paramount importance, and should be recorded immediately and accurately in the field. Accurate typographical and geological data should be included, stating as much as the collector knows about the localized stratigraphy and lithologies present. Record data on labels or bags used for collection. Avoid writing on specimens with any type of marker pen.

Each specimen should be individually wrapped. Pieces of specimens broken during extraction should be wrapped separately. Newspaper is too abrasive to use for packing minerals, and acid-free tissue, which is absorbent and easily removable from wet material, is preferable. The specimen should be wrapped so that any delicate areas are protected by the bulk of the wrapping.

Cloth bags with labels attached are ideal for the transportation of field samples. Delicate material, however, should be wrapped in acid-free tissue before placing in the cloth bag. Never use cotton wool. Figure AII.1 shows the disastrous result of using cotton wool for storing a fragile zeolite specimen.

It is practically impossible to transport minerals such as delicate zeolites, or brittle minerals such as wulfenite, for long distances in complete safety. The use of open tissue-lined trays and transport by hand is perhaps the only practical method.

Figure AII.1 Result of storing zeolite in cotton wool.

Appendix III

Bob King

Cleaning minerals

The term 'cleaning' implies the removal of foreign matter from mineral assemblages, i.e. removal of dust, dirt or extraneous clay – not the removal of associated minerals, including the products of oxidation, from the surface of the principal mineral(s). The latter is development, which often demands the use of chemical reagents that can promote instability. Before any cleaning process is begun, some insight into the basic chemistry of the material is essential, so that a technique that will not affect the mineral assemblage or paragenesis can be chosen.

Water is a chemical reagent and Table AIII.1 lists some of the minerals that are affected by contact with water. Cleaning may wash off loosely adhering minerals such as hydrated iron oxides, thereby destroying part of the paragenesis. The cleaning of water-sensitive minerals can in some cases be achieved by means of organic solvents, but is generally best done by gentle dry brushing or 'airbrasive' methods, using abrasives such as fine cork or powdered walnut shell.

Wet mineralogical material should be dried slowly, under controlled conditions. It should not be oven-dried, but should be allowed to dry in ambient air on acid-free absorbent tissue. Drying in an environmentally controlled chamber may be essential in some cases (but see Chapter 3). Rock specimens are generally more resilient, and can be dried in a low oven (40°C to 50°C maximum).

Some absorbent and porous minerals are listed in Table AIII.2.

Many mineral species occur in hair-like acicular or capillary habits, and their cleaning is technically quite difficult. Once exposed to the atmosphere, specimens such as acicular zeolites will quickly become dulled. Cleaning should only be attempted if absolutely essential, and then only dry-cleaning methods, such as the use of a photographer's hand blower, should be employed.

Well-cleaved minerals should not be submitted to long-term soaking in water, nor should they be cleaned with detergent solutions. Cleavages permit solutions to infiltrate deep into the material by capillary action. Gypsum crystals, for example, become opalescent after dipping in soap solutions, which infiltrate the [010] cleavages.

Well-cleaved minerals should not be immersed in or brushed with solvents such as ethanol or acetone. Such action may cause the formation of coloured films within the cleavages, and may physically weaken the material. If necessary, the only cleaning should be by gentle brushing or the use of 'airbrasives', with soft abrasives such as cork or powdered walnut shell.

Many clays may be removed from rock samples by simply soaking in water, with gentle brushing out of interstices where necessary. Harder clay may be loosened by means of a thin wooden splint. Do not use metal probes, needles or nylon brushes. The use of chemical deflocculants, and chelating agents such as sodium hexametaphosphate or EDTA and its salts, is not recommended, because the cleaning materials will tend to remove calcium, magnesium, iron, etc. Ultrasonic cleaning, which may be used on some rock specimens, is not generally recommended for minerals.

Clay removal from water-soluble minerals can sometimes be effected by the use of ethanol or propan-2-ol jetted from a wash bottle. To prevent caking of clay on the minerals' surface, the alcohol used should not be allowed to evaporate until all the clay particles have been removed.

Development techniques are used to a limited extent on some mineral assemblages, but few have

Table A III.1 A list of some minerals affected by water

		Rate of solubility*
Alum	$KA1(SO_4)_2.12H_2O$	1
Alunogen	$A1_2(SO_4)_3.18H_2O$	1
Amarantite	$Fe\ SO_4OH.3\frac{1}{2}H_2O$	Decomposed
Apthiatalite	$NaKSO_4$	1
Apjohnite (Glocker)	$MnA1_2(SO_4)_4.22H_2O$	1
Arsenolite	As_2O_3	3 in hot water
Autunite	$Ca(UO_2)_2(PO_4)_2.12H_2O$	1
Azurite (Beudant)	$Cu_3(CO_3)_2(OH)_2$	Decomposed by hot water
Beyrichite	Ni_3S_4	Decomposed by hot water
Bianchite	$(Zn,Fe)SO_4.6H_2O$	1
Bilinite	$Fe^{2+}Fe^{3+}_2(SO_4)_4.22H_2O$	1
Bischofite (of Ochsenius)	$MgC1_2.6H_2O$	1
Blödite	$Na_2Mg(SO_4)_2.4H_2O$	1
Borax	$Na_2B_4O_7.10H_2O$	1
Botryogen	$MgFe(SO_4)2OH.7H_2O$	3 in hot water
Boussingaultite	$(NH_4)_2Mg(SO_4)_2.6H_2O$	1
Burkeite	$Na_6SO_4(CO_3)_2$	1
Carnallite	$KMgC1_3.6H_2O$	1
Cerussite	$PbCO_3$	Decomposed by hot water
Chalcanthite	$CuSO_4.5H_2O$	1
Colemanite	$Ca_2B_6O_{11}.5H_2O$	3
Copiapite	$R^{2+}Fe^{3+}_4(SO_4)_6(OH)_2.nH_2O$	(1) where R includes Fe^{2+},Mg,A1, Cu or Na_2
Coquimbite	$Fe(SO_4)_3.9H_2O$	1
Cotunnite	$PbC1_2$	1
Cryolite	Na_3A1F_6	3
Cyanochroite	$K_2Cu(SO_4)_2.6H_2O$	1
Darapskite	$Na_3NO_3SO_4.H_2O$	1
Dietrichite	$(Zn,Fe,Mn)A1_2(SO_4)_4.22H_2O$	1
Dietzeïte	$Ca_2(IO_3)_2CrO_4$	2
Dolerophane	Cu_2SO_5	1
Douglasite	$K_2FeC1_4.2H_2O(?)$	3
Epsomite	$MgSO_4.7H_2O$	1
Erythrosiderite	$K_2FeC1_5.H_2O$	1
Ettringite	$Ca_6A1_2(SO_4)_3(OH)_{12}.26H_2O$	2
Fernandinite	$Ca(VO)_2V_{12}O_{28}.14H_2O$	2
Ferrinatrite	$Na_3Fe\ (SO_4)_3.3H_2O$	3
Gaylussite	$Na_2Ca(CO_3)_2.5H_2O$	3
Glauberite	$Na_2Ca(SO_4)_2$	2
Goslarite	$ZnSO_4.7H_2O$	1
Greenockite	CdS	Powdery varieties readily decomposed by hot water; crystallized varieties less so
Gypsum	$CaSO_4.2H_2O$	3
Halite	$NaC1$	1
Halotrichite (of Glocker)	$Fe\ A1_2(SO_4)_4.22H_2O$	1
Hanksite	$Na_{22}K(SO_4)_9(CO_3)_2C1$	1
Hewettite	$CaV_6O_{16}.9H_2O$	3
Hexahydrite	$MgSO_4.6H_2O$	1
Hieratite	K_2SiF_6	1
Hydrocyanite	$CuSO_4$	1
Ilesite	$(Mn,Zn,Fe)SO_4.4H_2O$	1
Inyoite	$Ca_2B_6O_{11}.13H_2O$	1 in hot water

Continued

Table A III.1 *Continued*

		Rate of solubility*
Kainite	$KMgSO_4C1.3H_2O$	Decomposed to epsomite and sylvite
Kaliborite	$KMg_2B_{11}O_{19}.15H_2O$	3
Kalinite	$KA1(SO_4)_2.11(?)H_2O$	1
Kernite	$Na_2B_4O_7.4H_2O$	2
Kieserite	$MgSO_4.H_2O$	2
Koenenite	$Mg_5A1_2C1_4(OH)_{12}.2(?)H_2O$	Decomposed to scaly precipitate
Kornelite	$Fe_2(SO_4)_3.7H_2O$	1
Krausite	$KFe (SO_4)_2.H_2O$	Slowly decomposed
Kremersite	$(K,NH_4)_2FeC1_5.H_2O$	1
Kröhnkite	$Na_2Cu(SO_4)_2.2H_2O$	1
Langbeinite	$K_2Mg_2(SO_4)_3$	3
Lanthanite	$(Ca,Ce)_2(CO_3)_3.9H_2O$	Decomposed by hot water
Larnite	Ca_2SiO_4	2
Lautarite	$Ca(IO_3)_2$	3
Lecontite	$Na(NH_4,K)SO_4.2H_2O$	1
Leonite	$K_2Mg(SO_4)_2.4H_2O$	1
Löweïte	$Na_2Mg(SO_4)_2.2.5H_2O$	1
Malachite	$Cu_2CO_3(OH)_2$	Decomposed by hot water
Mallardite	$MnSO_4.7H_2O$	1
Mascagnite	$(NH_4)_2SO_4$	1
Melanterite	$FeSO_4.7H_2O$	1
Mendozite	$NaA1(SO_4)_2.11(?)H_2O$	1
Metahewettite	$CaV_6O_{16}.9H_2O$	3
Metarossite	$CaV_2O_62H_2O$	2
Metavoltine (of Blaas)	$K_4Na_3Fe^{2+}Fe^{3+}_5(O_4)_{12}.16H_2O$	3
Millerite	NiS	Decomposed by hot water
Minasragrite	$(VO)_2H_2O(SO_4)_3.15H_2O$	1
Mirabilite	$Na_2SO_4.10H_2O$	1
Misenite	$K_8H_6(SO_4)_7$	1
Mitscherlichite	$K_2CuC1_4.2H_2O$	1
Morenosite	$NiSO_4.7H_2O$	1
Nahcolite	$NaHCO_3$	1
Natrochalcite	$NaCu_2(SO_4)_2OH.H_2O$	2
Natron	$Na_2CO_3.10H_2O$	1
Nitratine	$NaNO_3$	1
Nitre	KNO_3	1
Nitrobarite	$Ba(NO_3)_2$	1
Nitrocalcite	$Ca(NO_3)_2.nH_2O$	1
Nitromagnesite	$Mg(NO_3)_2.6H_2O$	1
Orpiment	As_2S_3	Soluble (3) in hot water
Oxammite	$(NH_4)_2C_2O_4.H_2O$	1
Pascoite	$Ca_2V_6O_{17}.11(?)H_2O$	2
Phillipite	$Cu_3Fe_2(SO_4)_6.40H_2O$	1
Pickeringite	$MgA1_2(SO_4)_4.22H_2O$	1
Picromerite	$K_2Mg(SO_4)_2.6H_2O$	1
Pintadoite	$Ca_2V_2O_7.9H_2O$	2
Pisanite	$(Fe,Cu)SO_4.7H_2O$	1
Polyhalite	$K_2MgCa_2(SO_4)_4.2H_2O$	Decomposed with separation of gypsum
Rinneïte	$K_3NaFeC1_6$	1
Römerite	$Fe^{2+}Fe^{3+}_2(SO_4)_4.12H_2O$	1
Rossite	$CaV_2O_6.4H_2O$	3
Sal-ammoniac	NH_4C1	1
Sassolite	$B(OH)_3$	1
Schairerite	$Na_3SO_4(F,C1)$	2
Schröckingerite	$NaCa_3UO_2SO_4(CO_3)_3F.10H_2O$	1
Sideronatrite	$Na_3Fe(SO_4)_3.3H_2O$	Soluble in hot water with decomposition
Siderotil	$FeSO_4.H_2O$	1

Sulfoborite	$Mg_3SO_4B_2O_5.4.5H_2O(?)$	Decomposed in water
Sulphohalite	$Na_6(SO_4)_2ClF$	3
Sylvine	KCl	1
Syngenite	$K_2Ca(SO_4)_2.H_2O$	3 with separation of gypsum
Szomolnokite	$FeSO_4.H_2O$	2 with formation of brown solution
Tachhydrite	$CaMg_2Cl_6.12H_2O$	1
Taylorite (of Dana)	$(K,NH_4)_2SO_4$	1
Teschemacherite	$(NH_4)HCO_3$	1
Thenardite	Na_2SO_4	2
Thermonatrite	$Na_2CO_3.H_2O$	1
Tincalconite	$Na_2B_4O_7.5H_2O$	1
Trona	$Na_3H(CO_3)_2.2H_2O$	1
Tschermigite	$NH_4Al(SO_4)_2.12H_2O$	1
Ulexite	$NaCaB_5O_9.8H_2O$	2
Vanthoffite	$Na_6Mg(SO_4)_4$	2
Villiaumite	NaF	1
Voltaite	$HK_2Fe^2_4(Fe,Al)_3(SO_4)_{10}.13H_2O$	Decomposes with yellow precipitate
Zinc-copper-melanterite	$(Zn,Cu)SO_4.7H_2O$	1

*Rate of solubility
1 Strongly affected.
2 Moderately affected.
3 Slightly or slowly affected.

any positive scientific benefit. Each mineral specimen is unique, and it is therefore essential that it should retain its full scientific value and potential after the development process has been completed. Hansen[1] details the arguments against the use of chemical and physical development techniques; these are summarized below:

1 The destruction of a natural association of mineral species by the removal of one or more associates results in the loss of scientific interest in that association.
2 The unnatural appearance following the use of chemicals results in the loss of scientific and aesthetic value.
3 There can be no standard application of a development technique. No two mineral associations respond to chemicals in quite the same way.
4 The use of chemicals may induce instability.
5 Incomplete removal of the chemicals used leads to additional short- or long-term risk to the minerals treated.

For certain classes of minerals, development may not represent a significant threat. However, it may be necessary to develop a specimen for scientific examination. Whenever it is decided to use a chemical technique the following criteria should be adopted:
● Use the mildest techniques available
● If several specimens are available, always use the least important first. If only one high-quality specimen is available, then do not use any development techniques.
● Always record precisely what has been done. This data should either accompany the developed specimen or be entered in the collection register.

Films of iron oxides are commonly found on minerals, either as a result of oxidation or paragenesis. Removal of these by various acid-based techniques is potentially destructive and should be avoided. Waller[2] has improved on a technique described by Mehra and Jackson,[3] based on the use of anions which sequester ferrous ions. The method uses sodium dithionite to reduce ferric to ferrous ions, and sodium citrate to chelate, or sequester ferrous ions and sodium bicarbonate as a buffer to maintain neutrality.

There are limitations to the use of such methods: calcium or magnesium minerals such as dolomite, calcite and aragonite, for example, should not be treated with chelating agents. The method is best reserved for quartz and silicates. To quote from Davidson's paper[4] '. . .the danger of these operations may not be apparent to the non-specialist. . . A group of minerals which has undergone acid treatment in the museum laboratory is no longer truly representative of a natural occurrence, and its scientific value is less than that of a similar, but untreated specimen'.

Table A III.2 Some absorbent and porous minerals

Aluminite	$Al_2SO_4(OH)_4.7H_2O$
Chloropal	Silicate of Fe^3
Chrysocolla	$CuSiO_3.2H_2O$
Cimolite	$Al_4Si_9O_{24}.6H_2O$
Gibbsite (of Torrey)	$Al(OH)_3$
Halloysite (of Berthier)	$Al_2Si_2O_5(OH)_4.2H_2O$
Hydrozincite	$Zn_5(CO_3)_2(OH)_6$
Illite	$K_{2-3}Al_{11}Si_{12-13}O_{35-36}(OH)_{12-13}$
Limonite	Hydrated oxide of Fe^3
Kaolinite Group { Nacrite / Dickite }	$Al_2Si_2O_5(OH)_4$
Metarossite	$CaV_2O_6.2H_2O$
Montmorillonite	$R_{-0.33}(Al,Mg)_2Si_4O_{10}(OH)_2.nH_2O$, where R. includes Na., K., Mg_2, Ca_2
Opal	$SiO_2.nH_2O$
Palygorskite	$(Mg,Al)_5(Si,Al)_8O_{20}(OH)_2.8H_2O$
Pyrolusite	MnO_2
Sal-ammoniac	NH_4Cl
Saponite (of Svanberg)	Aluminosilicate of Mg
Sassolite	$B(OH)_3$
Scarbroite	A clay mineral containing c.44% Al_2O_3 and c.6% SiO_2
Searlesite	$NaBSi_2O_6.H_2O$
Sepiolite	$Mg_3Si_4O_{11}.5H_2O$
Sideronatrite	$Na_2Fe^3(SO_4)_2OH.3H_2O$
Turquoise	$CuAl_6(PO_4)_4(OH)_8.5H_2O$
Valentinite	Sb_2O_3

References

1 HANSEN, M., 'Cleaning delicate minerals', *Mineralog. Record*, 15, p. 103 (1984)
2 WALLER, R.R., 'A rust removal method for mineral specimens', *Mineralog. Record*, 11, pp. 109-10 (1980)
3 MEHRA, O.P. and JACKSON, M.L., 'Iron oxide removal from soils and clays by a dithionite–citrate system buffered by sodium bicarbonate', in Swineford, A. (Ed.), *Mon. No. 5, Earth Sciences Ser.* Pergamon Press, pp. 317-27 (1958)
4 DAVIDSON, C.F., 'Acid treatment of rocks and minerals', *Museums J.*, 42, p. 292 (1942)

Appendix IV

Bob King

Repair and consolidation of minerals and rocks

Damaged or broken minerals, although of little aesthetic value, are almost always of scientific value. If repair is necessary, care should be taken in the selection of adhesives. The so-called 'permanent' adhesives, such as polyesters, cyanoacrylates and epoxy resins, should not be used for the repair of minerals. The repair will often be stronger than the mineral, and the adhesive used may react unfavourably with the specimen. Refer to Table AIV.1 for further information. Details of any repair should form an integral part of the specimen's records.

Consolidation of friable rocks and mineral assem-
blages may be unavoidable, for example, shales, mudstones, some loosely cemented sandstones and groups of crystals on substrates which would not survive handling, transport or long-term storage. Deep penetration by consolidants may be achieved by means of vacuum impregnation, but degassing of the volatile carrier solvent within the specimen can cause serious structural damage. Airbrush spraying of consolidants is not recommended, as it is difficult to control coverage. For detailed information on consolidation methods and treatment of old conserved material refer to texts such as Rixon,[1] Croucher and Woolley[2] and Brommelle *et al.*[3] Minerals should not be consolidated, as their physical characteristics could be altered and later analyses jeopardized.

Table AIV.1 Some consolidants and adhesives for use on groundmass, matrix and rock specimens

Name and supplier	Constituents	Application
Paraloid B series (Rohm and Hass)	Ethyl methacrylate – methyl methacrylate copolymers	General cosolidant, penetration often better than PVA, less tack
Dri-Film 104 (General Electric)	Pre-polymerized methylalkyl-siloxane	Nor for use on siliceous material: difficult to remove later
Primal AC range (Rohm and Hass)	Acrylic emulsions and colloidal dispersions	Limited use as consolidant for geological material
Polyvinyl acetate emulsions, PVA (several formulations, e.g. Gelva, Vinamul Tenaxatex, etc.)	Poly(vinyl acetate) plasticizer(s), water stabilizer(s)	General adhesive and consolidant. High solids content, useful for very porous material.[1] Most emulsions are slightly acidic, so could corrode some calcareous material. Good on waterlogged material[1]
PVDC, Saran	Poly(vinylidene chloride)	Excellent consolidant where low permeability to water vapour required May discolour
Butvar B98/B90/B76 (Monsanto)	Poly(vinyl butyral)	General purpose[3, 4]
Dow Corning (Silanes (Z and T series) (Dow Chemical)	Methyl trimethoxy-silane	Not for use on siliceous material: difficult to remove later

Table AIV.1 *continued*

Name and supplier	Constituents	Application
Polystyrene (BDH)	Polystyrene, various mol. wts	Brittle, limited use
Tensol, Perspex (ICI)	Poly(methyl methacrylate)	Good for temporary repairs to lightweight material[5]
Lepages cement (H.C. Stephens Ltd)	Cellulose acetate Cellulose acetate/butyrate	Extensively used for repair of minerals formerly
HMG, Durafix	Cellulose nitrate	Should not be used in conservation, unstable, may be spontaneously inflammable
Polyvinyl alcohol (BDH)	Polyvinyl alcohols (various mol. wts)	Useful as water-based consolidants for paper, etc.
Carbowax range (Union Carbide)	Polyethylene glycols of various mol. wts	Mid-range mol. wt. pegs (1,500–6,000) are useful for temporary support for specimens
Araldite range (CIBA Geigy)	Epoxide + (eg) amine catalyst	Limited use in geological specimen conservation
Powerbond, Loctite	Cyanoacrylate monomers catalyzed by moisture	Used in fossil conservation as adhesive; difficult to reverse. Not recommended for repair of mineral specimens (note colour reactions with certain matrices, eg. blue with iron-rich types)

References

1 RIXON, A.E., *Fossil Animal Remains: Their preparation and conservation*, London (1976)
2 CROUCHER, R. and WOOLLEY, A.R., '*Fossils, minerals and rocks. Collection and preservation*', British Museum, Natural History, London (1982)
3 BROMMELLE, N.S., PYE, E.M., SMITH, P. and THOMSON, G. (Eds), 'Adhesives and Consolidants', in *Preprint of Contributions to Paris Congress, International Institute for Conservation* London (1984)

Index